DEVELOPMENT AND DEMOCRACY

Studies in Critical Social Sciences Book Series

Haymarket Books is proud to be working with Brill Academic Publishers (www.brill.nl) to republish the *Studies in Critical Social Sciences* book series in paperback editions. This peer-reviewed book series offers insights into our current reality by exploring the content and consequences of power relationships under capitalism, and by considering the spaces of opposition and resistance to these changes that have been defining our new age. Our full catalog of *SCSS* volumes can be viewed at https://www.haymarketbooks .org/series_collections/4-studies-in-critical-social-sciences.

DEVELOPMENT AND DEMOCRACY

Relations in Conflict

EDITED BY

VÍCTOR MANUEL FIGUEROA SEPÚLVEDA

Haymarket Books
Chicago, IL

First published in 2017 by Brill Academic Publishers, The Netherlands.
© 2017 Koninklijke Brill NV, Leiden, The Netherlands

Published in paperback in 2018 by
Haymarket Books
P.O. Box 180165
Chicago, IL 60618
773-583-7884
www.haymarketbooks.org

ISBN: 978-1-60846-088-5

Trade distribution:
In the U.S. through Consortium Book Sales, www.cbsd.com
In the UK, Turnaround Publisher Services, www.turnaround-uk.com
In Canada, Publishers Group Canada, www.pgcbooks.ca
All other countries, Ingram Publisher Services International, ips_intlsales@
ingramcontent.com

Cover design by Jamie Kerry and Ragina Johnson.

This book was published with the generous support of Lannan Foundation
and the Wallace Action Fund.

Printed in United States.

10 9 8 7 6 5 4 3 2 1

Library of Congress Cataloging-in-Publication Data is available.

Contents

Acknowledgements

The editor of this volume along with all of its contributors wish to thank the National Council of Science and Technology (CONACYT) of Mexico for funding the research project "Science, Development and Democracy."

List of Tables and Figures

Tables

Figures

Notes on Contributors

Irma Lorena Acosta Reveles
is a political scientist and researcher at the Autonomous University of Zacatecas. Since 2010, she has been a member of Mexico's National System of Researchers (CONACYT) and served as director of the Doctoral Program in Political Science for the 2013–2016 period. Her published research has investigated problems of rural society and the agrarian economy of Mexico and Latin America in addition to labour issues and the conditions of academic work.

Leonel Álvarez Yáñez
was professor and researcher at the Department of Political Science at the Autonomous University of Zacatecas where he served as director until mid-2015. His research grappled with issues concerning science, development, governance, political regimes and governability. Among other works, he authored La democratización figurada: Ingobernabilidad causada por el impacto socio-político de las reformas estructurales en México 1985–2006 (UAZ 2008). Dr. Álvarez passed away on 7 July 2015.

Jesús Becerra Villegas
is a researcher at the Autonomous University of Zacatecas, Mexico. His early academic production dealt with language, narrative semiotics and the media. More recently, his interests have included problems of capitalism in crisis, regulation and institutional learning. He is author of El orden de la comunicación (UAZ, 2009) and the essay "Appropriating financial crisis: a communication studies reading."

Ximena de la Barra
was adjunct associate professor in Columbia University in the field of social policy and urban planning before taking up a distinguished career in the United Nations where she worked first in UN Habitat and later in UNICEF as a Global Advisor for Public Policy. Retired, she is now a development consultant and researcher in various areas related to political and social change. She is editor of Neoliberalism's Fractured Showcase: Another Chile is Possible (Brill 2011) and co-author of Latin America after the Neoliberal Debacle: Another Region is Possible (Rowman and Littlefield, 2009).

Héctor de la Fuente Limón
is a professor and researcher at the Department of Political Science at the Autonomous University of Zacatecas. He has specialized in the study of problems in development and its impact on social change in Latin America. He is a member of the Mexican National System of Researchers (CONACYT).

R.A. Dello Buono
is professor of Sociology at Manhattan College in New York City and was an invited lecturer at the Autonomous University of Zacatecas. His research areas are in comparative social problems and the political economy of development with an area focus in Latin America. He is co-editor of América Latina: Alternativas para el Desarrollo (FLACSO-Cuba, 2013) and co-author of Latin America after the Neoliberal Debacle: Another Region is Possible (Rowman and Littlefield, 2009).

Sergio Octavio Contreras
is professor at the universities of La Salle Bajío, Autónoma de Durango, Tecnológico de Monterrey and Pedagógica Nacional. His current research is on the political use of the internet and social media networks. He is author of Medios de Comunicación y Elecciones (2011) and Reproducción, Crisis, Organización y Resistencia (BUAP, 2014).

Silvana Andrea Figueroa Delgado
is professor and Director of the Political Science Department at the Autonomous University of Zacatecas. Her areas of research include the relationship of universities with science, technology and development. She is author of El Estado y el trabajo científico en el desarrollo (BUAP, 2009) and La ciencia y tecnología en el desarrollo: Una visión desde América Latina (with Germán Sánchez Daza and Alejandra Vidales Carmona) (UAZ, 2008).

Víctor Manuel Figueroa Sepúlveda
is professor and researcher at the Graduate Department of Political Science at the Autonomous University of Zacatecas, Mexico. His main research interests revolve around political and socio-economic problems of Latin America. He is author of Reinterpretando el subdesarrollo. Trabajo, clase y fuerza productiva en América Latina (Siglo XXI, 1986) and Industrial Colonialism in Latin America: The Third Stage, (Brill, 2013). He is presently working on a project concerning democracy and technological development in the United States, 1970–2015.

Ernesto Menchaca Arredondo

is professor and researcher in the Department of Political Science at the Autonomous University of Zacatecas. His areas of research include political thought and contemporary social processes. Among his recent publications are: "Vulnerabilidad municipal en el estado de Zacatecas, México" and "Nuevas rutas sobre el bienestar y las condiciones de vida de la población: la situación de la ciudad de Zacatecas."

Miguel Omar Muñoz Domínguez

is a political scientist at the Autonomous University of Zacatecas who works in the Office of Evaluation and Institutional Information and teaches in the master's program in Educational Information Technology in the Department of Higher Education. His research interests include the linkages between enterprises and universities with respect to their joint collaboration in the development of patents, science and technology.

Alexandre M. Quaresma de Moura

is a writer and researcher on the anthropological and socio-ecological impact of technology. He has a special interest in the critique of technology and is currently exploring controversies regarding artificial intelligence. He is author of Humano-Pós-Humano-Bioética, conflitos e dilemas da Pós-modernidade (Common Ground Publishing, 2014); Engenharia genética e suas implicações (ed.) (Madrid, 2014); and Nanotecnologias: Zênite ou Nadir? (Escriba, 2010).

Cristina Recéndez Guerrero

is professor of sociology and researcher at the Department of Political Science at the Autonomous University of Zacatecas. She is a specialist in educational policy and management. Among her current interests is the study of labour transitions. She is co-author of De la contrarreforma universitaria neoliberal a la resistencia en América Latina (Elalpeh, 2009); co-editor of Políticas educativas y universidad pública (UAZ, 2010); and co-author of Relación laboral y cambio en la reproducción familiar (ZZBalsa, 2014).

Introduction

The research project "Science, Development and Democracy" that ultimately resulted in the present volume was carried out with the support of the National Council of Science and Technology of Mexico. As part of that project, a group of researchers convened an international colloquium "Development and Democracy: Present Challenges" in June 2015. The event took place in the context of a critical review that had been developed some time earlier during the formulation of the study. A preceding colloquium held a year earlier had already borne witness to a certain shift in some of the initial intuitions that guided the research.

Originally, this team research project was conceived with the following suppositions:

- Latin American societies did not create an internal division of labour that included general (scientific) labour where the development of productive forces takes place. As a consequence, technological progress has been monopolized by the developed countries which are defined as such precisely because they have created such a division, i.e., one that separates scientific labour from immediate labour. Underdeveloped regions share an asymmetrical integration with the developed world in which the latter manages to define the direction and rhythm of growth within the former.
- In the area of international relations, increasing integration generates a tendency towards deficits in the balance of trade, an insatiable need for credit and investments, a structural inclination towards indebtedness, and an equally constant flow of remittances in the form of profits and interest. In short, these disequilibria make it impossible to achieve a relatively balanced and economically stable process of development.
- On the domestic plane, underdeveloped countries display distinctive characteristics that include the importation of progress and the outflow of resources abroad (that could otherwise be used locally) while at the same time creating elevated surpluses of population that become excluded from the main economy. This places narrow limits on democracy just as it does on the distribution of wellbeing and the exercise of political liberties. The social discontent that arises in such conditions tends to be unceasing and is frequently met with repression.

Overcoming the situations embodied within the above suppositions within the context of global capitalism is not impossible as the experience of countries such as China and South Korea seems to indicate. Nonetheless, the deepening of democracy seems to be frustrated in both of these cases, although perhaps to a lesser extent in South Korea.

Socialism as conceived of by the founders of Marxism (Marx and Engels) cannot be realized under the conditions of underdevelopment. Workers cannot control what they have not created, that is, the means of production of general labour, with all that this implies in terms of the tendencies alluded to above. This will remain the case unless socialism can be installed in one or several developed countries such that it could make possible a solidarity-based international organization of labour and the construction (or beginning) of a new world order on a global scale.

Peoples living in underdevelopment will continue struggling for popular governments and their struggles should be supported to the end of opening and expanding democratic spaces that can make possible a growing participation of the popular sectors in the organization of their societies. This in turn can provide preparatory lessons for advancing towards larger ends such as the promotion of capacities that are needed for the endogenous production of progress. The appropriation of these capacities by workers will put them in an objectively better position.

The rise of progressive governments in Latin America over recent decades generated new expectations for advancing material development and democratic decision making. While their democratic aspirations have been reasonably met, their ability to advance the domestic productive forces has proven to be significantly less impressive. It is certainly the case that they have been constrained by foreign economic and political harassment, but that is precisely something that cannot be avoided so long as they remain unable to produce progress locally. What is needed above all is progress that can neatly interface with the economic presence of popular masses engaged in political participation.

In recent years, the global economic and political situation has become significantly more complicated. The developed countries have been unable to overcome the tendency for low rates of growth that began to set in at the end of the 1970s. The conditions of unemployment and social inequality have spilled over as has poverty. Meanwhile, authoritarianism has extended its reach while liberal democratic liberties have experienced serious repression. A new world capitalist order can be seen evolving where the powerful are ruling amidst highly conflictual relations. Capitalism has failed to develop the means to manage economic and technological development in a democratic manner. At the same time, the activism of Latin America with the exception of Brazil in certain aspects remains practically irrelevant.

The works contained in this volume serve to analyse the conflicting relations of technological development and democracy that are unfolding in this new and changing environment. The collection begins with a stimulating philosophical and historical reflection by Alexandre Quaresma about the significance of techno-science in his work entitled "A Critique of the Origin and Foundations of the New Inequality among Men." Following a brief historical synopsis, he describes some of the impressive technological-scientific developments of recent decades and their contradictory relationship with social inequalities. His critical work rages against the private appropriation of knowledge, the morale that reinforces it, and the resulting control that is imposed over societies, people and nature. The reflections offered by Quaresma serve to alert us to the consequences of technological knowledge that is subjected to the logic of profit, with particular emphasis on the effects it has on the exercise of democracy.

The next work by Víctor Manuel Figueroa entitled "Unemployment, Inequality and Technological Development" reflects on the structural decadence of capitalism in its present form. The notions of "technological unemployment" and "technological inequality" in their current magnitudes are presented as expressions of the material development of the new era that confronts us. Figueroa's object of inquiry is the United States and he proposes a critical reading of that country's developmental processes by exploring the new relations that are emerging between the worker and the means of production, highlighting those which are appearing between scientific work and smart machinery. His work situates the observable changes in a historical perspective, allowing him to analyse the effects that this economic process has upon the political organization of society.

In the following chapter entitled "Technology and Subsumption by Capital," Jesús Becerra picks up a discussion first suggested in the draft Chapter VI of *Capital* where Marx distinguishes between the forms of subsumption of labour in capital within historical stages of capitalism. With the aim of characterizing the conditions of the present capitalist crisis being experienced, Becerra's point of departure rests in drawing a distinction between the logical and historical dimensions of dialectical materialist thought in order to characterise the fundamental elements of the current social formation. Becerra formulates the structural conflicts of social densification and examines the uncertainty of the distinct types of actors that must confront them with their own strategies. He proposes to add to the formal and real forms of subsumption, a third form of a symbolic nature that emerges out of the first two industrial revolutions and becomes consolidated during the third. Based on this reading, symbolic domination operates as a mechanism by which capital can extend its logic beyond the processes of production, allowing it to appropriate the effects of social densification and uncertainty. Becerra identifies a variety of symbolic and technical resources that lends support to capitalism in the face of social conflicts of diverse types, all in order to subsume the world to its needs of reproduction.

In "The State and Freedom in Public Network Space," Sergio Octavio Contreras discusses the evolution of the relations of political power and the global network. He maintains that in the field of liberal democracy, people's use of internet as an open space for the exchange of content offers communicative freedom to exercise their right to freely express themselves. This faculty has become converted into a kind of social power that operates as a counterweight to political power and serves as a means to politicize people. Since the internet was freed up in 1993 for social use, however, it has confronted a series of threats in the form of various power structures. Following a review of registered cases over a four year period, Contreras determines that three forms of network regulation exist within current political regimes, namely, meanings control, market control and police control. Through the implementation of tools of legal/illegal controls, states acting in complicity with economic powers have constructed virtual dikes in order to contain network freedoms. Contreras offers the reader an introduction to internet freedoms from a technological perspective in his exploration of the forms of established power that seek to impose themselves upon informational networks.

Silvana Andrea Figueroa's work "Grey Areas in China's Growth: A Questionable Development" brings to light the key factors that explain the spectacular growth of the Chinese economy. She places special emphasis on Chinese strategies adopted to strengthen the scientific-technological apparatus, allowing it to reach a robust critical mass. As the importance of spending in research and development is widely known, Figueroa elucidates contrasting images in the relationship between this spending, the impressive economic growth of the country, and the questionable presence of development and democracy. These contrasts notwithstanding, she demonstrates that they act together with contemporary public policy to favour the construction of skills and the promotion of knowledge that become rooted in other political and social aspects that implicitly contribute to additional negative features in the established model of growth. In so doing, Figueroa questions the real extent of development and democracy in China and points to the massive contradictions existent in the economic role currently being played by this nation in the global scenario.

Cristina Recéndez in her "Economic Growth, Democracy and the Construction of Citizenship in South Korea" points to another emblematic country with notable economic advances. Her study seeks to define the context in which changes have taken place during the last six decades and to determine the factors that have contributed to the economic, political and social development of that country. She highlights in this work the role played by the state in a global context of neoliberalism and a local context of the struggle for democracy. Recéndez also discusses experiences aimed at overcoming underdevelopment, an issue of key importance for Latin America, and this effectively serves to dig deeper into the relations between economic advances and political democracy in a transformational context.

In "Latin American Democracy as an Alternative Work in Progress," R.A. Dello Buono and Ximena de la Barra argue that Twentieth Century attempts to achieve development brought liberal democratic states up against their limits. Keynesian strategies required a strong state that systematically promoted reforms but capitalist expansion ultimately produced an insurmountable crisis in the import substitution model. As the neoliberal strategy supplanted Keynesian formulas so that global capital could extend and deepen its penetration, global superpower hegemony was restored but then also fell into multiple crises by 2008, thus leading to a marked moderation of neoliberal orthodoxy. At the same time, Latin America's electoral democracies, continually limited by foreign interventionism, have demonstrated that they are lacking for progressive programmes. Dello Buono and de la Barra argue that it is necessary to recover the sovereignty of national states and construct a regional sovereignty that articulates genuine democracy on behalf of the immense majority. There is a need to reconstruct the capacity of states to intervene in economic and social life in order to reduce unequal development. The aim is to develop democratic proposals that defend viable strategies for social transformation, beyond the electoral strategies of a so-called 'representative democracy.'

In his chapter "Acquiring Technology in the Mexican Private Sector: A Disarticulated 'Linkage' of the Triple Helix," Miguel Omar Muñoz examines Mexico's institutional weaknesses for producing domestic technology. He first describes the organizational cooperation model of the triple helix in which there is a favourable union of government, institutions of higher learning, and the private business sector that promotes the generation of technology and innovation in support of economic development and national productivity. He then counterposes this ideal model to the real form in which productive technologies are acquired by the Mexican business sector. Muñoz maintains that the impediments to the realization of this ideal model are to be found in the economic system in which developed countries technologically colonize the underdeveloped ones, a situation that neither the Mexican government nor Mexican business elites have demonstrated much interest in resolving. The result is a relationship of technological dependency where the actors involved in the processes of generating science and technology work together to conserve a backwards status quo that sustains the disarticulation of scientific and immediate labour. Muñoz supports his analysis with empirical data based on surveys conducted in Mexico.

In her chapter "Proliferation of the Corporate Agro-Industrial Model in Latin America," Lorena Acosta contributes to our knowledge regarding the effects, potential and socioeconomic, of the political and environmental challenges of science and applied technology in the agricultural and livestock sector. Recognising the coexistence of a capitalist and peasant economy in the region, Acosta focuses on the capitalist pole in agriculture since that is where technological resources are concentrated in the exploitation of the land. However, her analysis of the agricultural reality remains well connected to the interaction established with that sector's human inhabitants. Her emphasis on the conduct of the agro-industrial and foodstuffs corporations is based on the fact that the governments of the region view them as the most promising means to achieve prosperity. It is likewise the case that this corporate sector sets the value and prices that operate in the global market of agrarian goods. Acosta offers historical observations on the social changes linked to the diffusion of new agrarian technologies as well as a theoretical reflection that places the defining elements of the model into view. In two problematic fields, she systematises the active conflicts that sustain this extremely contradictory but profitable manner of producing.

"Well-Being and Happiness: Conditions for a New Conception of Development?" by Ernesto Menchaca Arredondo and Leonel Álvarez Yáñez sheds light on the points of reference in which Mexicans define their subjective well-being. Based on an explication of their own conception of well-being, they opt to include new subjective indicators that complement and express societal progress in a more insightful manner. Their research observes an increased perception among Mexicans of the importance of non-material factors such as the social family relations, the importance of social autonomy, and a very high value being placed on people's emotional lives. The authors maintain that these aspects constitute a conglomerate of subjective elements that must be taken into account, especially for a better characterisation of the exercise of democracy. In order to identify these principal components, a multivariate analysis was carried out so as to then situate them in network maps where it could be graphically illustrated as to how people describe their subjective well-being as well as their integration or exclusion that underlies the exercise of rights on the part of Mexican citizens.

Finally, Héctor de la Fuente in his "The Challenges of Democracy in Mexico" analyses the prevailing conditions and challenges that confront democracy in

Mexico, taking as his point of departure the neoliberal pattern of accumulation imposed over recent years. He emphasises the generation of an economy with lower rates of growth than those needed to sustain the growing size of the working-age population; the structural incapacity of the system to generate employment; the abrupt and continual fall of real wages; the increased precariousness of work; the increase in social inequality and economic exclusion; the increase of informal employment and migration as survival strategies; and the growth of the illegal economy and social violence. De la Fuente warns us that under these conditions, an accumulating mass of uncertainties and social conflicts have conspired to restrict the broader and equal participation of Mexican citizens in public issues, thereby undermining the autonomous forms of collective organisation and restricting the rule of law. The type of relations that have become established between the state and society have grown to be clientelist, vertical and ever more violent and as these relations are reproduced, they arc back to the state. The crisis of legitimacy among political parties and the main democratic institutions demonstrates a clear distancing of the system from citizens, providing further evidence of the limitations that underdevelopment imposes upon democracy.

Taking the collection as a whole, it becomes clear that development and democracy presently constitute relations in crisis whose reconciliation will surely prove elusive within the limits of capitalist organisation as we presently know it. The global capitalist reorganization that is currently taking place stands to pose ever deeper and more fundamental threats for the enjoyment of democracy.

Víctor Manuel Figueroa Sepúlveda
Autonomous University of Zacatecas, Mexico

CHAPTER 1

A Critique of the Origin and Foundations of the New Inequality among Men[1]

Alexandre M. Quaresma de Moura

Introduction

As the title of this article suggests, I intend to build off the engaging essay submitted by humanist philosopher Jean-Jacques Rousseau in response to an essay contest sponsored by the Dijon Academy in 18th Century France. Rousseau's brilliant and inspirational work remains poignant up to the present day, just as it was back in Rousseau's Age of Enlightenment. There has existed a deeply serious concern about the problems, conflicts and cleavages posed by the extreme concentration of power and wealth. This is especially acute in our present crisis laden era in view of the social and technological divergences that we are experiencing. There was a disconcerting contrast in Rousseau's time between the well-to-do and so-called nobility, and the excluded less fortunate, or rabble, as they were known. In our own day the disconcerting contrast persists between *les misérables* and the multimillionaires who produce transnational capital in its globalised phase. This will be used as a basis of comparison with the new technologies of the present whose commercial results, analogously, remain limited to small, elitist groups.

Technical practices are proving capable of deeply intervening in areas ever more determinant[2] in our lives and material world. In other times, we saw these

---

1   Of course the most correct term to use here would be "Human Beings" instead of "Men." Nevertheless, we prefer for historical, bibliographical and aesthetic reasons to retain the term "Men" in order to affirm our inspiration from the original title by Jean-Jacques Rousseau: *Discourse Upon the Origin and the Foundations of the Inequality Among Men*. The designation of "Critique" for "Discourse" is made because similarly "discourse" is one form used to construct a critique. And finally the term "New" serves to demarcate the core of our focus in terms of the temporal parameters of our analytical study.

2   The concept of *technological determinism* is based on the supposition that technologies have an autonomous functional logic that can be explained without societal reference. It presumes that technology is social only with respect to the purpose it serves and its aims in the mind of the observer. Technology in this sense resembles science and mathematics for its

from a vantage point of impotence and as obstacles to be overcome due to igno-rance or the lack of ingenuity, knowledge and appropriate machinery. Today, however, they are beholden as potent techno-sciences. It is important to men-tion that these technologies are available only to those who can afford to pay for them. Not to mention that the capability exists to generate new and more brutal forms of exclusion with unforeseen and unintended consequences, including the ability to scientifically generate aspects of the human being and other life forms in the biomolecular mould, i.e., in the human genome that can make pos-sible new forms of social exclusion and discrimination such as we have never seen before. It should be remembered that societies systematically produce technologies and that these in effect repeatedly transform those same societies in accordance with what we call technification.[3] This is evidenced today by way of neuronal implants to correct dysfunctions or increase performance; through probes and special telescopes to explore the cosmos; or interconnected online social networks; or in parallel virtual realities where we confuse ourselves with our telemetric avatars; or even with *intelligent evolutionary algorithms*,[4] con-trolling our flows of data, information and logistics with nanochips monitor-ing and tracking people and objects; with the production or bio-impression

---

apparent intrinsic independence from the social world. However, in contrast to science and mathematics, technology has immediate and powerful social impacts (Feenberg 2010: 72).

3 Technification: "From the perspective of experimental natural sciences, technification of human nature is simply the continuation of the well-known tendency to progressively make available the natural habitat. From the perspective of the lifeworld, however, our attitude varies once technification crosses the border that separates 'external' from 'internal' human nature" (Habermas 2002: 37).

4 Evolutionary algorithms are mathematical sequences that are governed by computer soft-ware of the same name that are conceived of in the logic of the natural biological evolution of life and nature. The general idea is the following: mechanistic entities are placed in a vir-tual environment composed of the high and low points of an artificially simulated relief, in a strict and specific regime of competition, with the object being to discover new routines that can allow for traversing paths that lead to the most elevated points within the defined environment. Those that manage to reach their objectives, such as occurs with life forms in evolution, are rewarded with survival, while analogously those that fail are excluded as they die and are removed from the competition since they cease to exist. The complex processes of natural selection replicate the competitive environment of natural biological evolution and, with the help of fast and powerful computers, the scientists of today that research these soft-ware creations can simulate millennia in short order. This contrasts with the time consumed by natural biological evolution, since continually repeating this same operation of competi-tion and adaptation enables the rise of a kind of singular "intelligence" that confers to this machinery a highly sophisticated quality of being able to learn how to learn from its own errors and advances.

of complete human organs, cloning, neo-eugenics; or with live and intelligent computers connected to our bodies; and with implants and artificial prostheses of all kinds incorporated into ourselves and our culture. All of this leads us to ask: have we transformed ourselves into something else that is no longer human? Or on the contrary, have we technologically created part of what we are, i.e., realized the nature of what we are and of what has always moved us and continues to move us through time?

The Ancient Problem of Inequality

The French Academy of Dijon publicly put forward a problem in 1754: "What is the origin of inequality among men? Is it authorised by natural law?" Jean-Jacques Rousseau must have felt particularly inspired by the question as he left Paris and took refuge in the forests of Saint-Germain in search of reflection and inspiration. At that moment in a bygone era, the rustic countryside was no longer seen as the threat it once was, where primitive humans fought with the animals for their existence, but rather as an ally where Rousseau could seek inspiration to write his celebrated *Discourse Upon the Origin and the Foundations of the Inequality Among Men*. He knew perfectly well that human beings were now ever more frightful than the natural world that surrounded them. Between the Noble Savage idealized by French writers and the advances, developments, and progress regarded by Rousseau as "promiscuities" of modern Paris that were boiling in the 18th Century, there was an enormous gap. It was there that he believed rested the origin of this flagrant inequality that has long afflicted humanity. That was Enlightenment thought being born while naive and terrible vestiges of bestial ignorance and myth dissipated, giving way to an enlightened new era of clarification and knowledge that was truly significant for the human being.

Rousseau in his response was more strange and paradoxical than it would at first appear, for in his Discourse, he blames precisely whoever formulated the question. That was the very human knowledge and progress that has accumulated since time immemorial, represented in actions by the Scientific Academy of Dijon, by the state, and in the final instance by the social contract and similarly all of the social, institutional and cultural behavioural derivations that could occur with that indeterminate being. As Sloterdijk (2000: 34) asserts, human knowledge cleverly and gregariously organises itself for survival: "Failing as an animal, this indeterminate being rushes out of its environment and proceeds to conquer the world in an ontological sense." Thus, the human being masters fire and carves out of stone in order to make primitive tools and armaments. Such epic events have had a profoundly dramatic impact upon primitive human cultures, and in this way, these early and innovative techniques initiated a great technological adventure that would continue into the present time with its overpowering microbiology technologies. In that odd time, twenty-five decades prior to ours, Rousseau was already acknowledging something that had been hidden for the future of humanity, and for that reason he reflected in a timid manner:

> There is, I feel, an age at which the individual man would wish to stop: you are about to inquire about the age at which you would have liked your whole species to stand still. Discontented with your present state, for reasons which threaten your unfortunate descendants with still greater discontent, you will perhaps wish it were in your power to go back; and this feeling should be a panegyric on your first ancestors, a criticism of your contemporaries, and a terror to the unfortunates who will come after you.
>
> ROUSSEAU 2005: 80

What Rousseau refers to here as that 'age' at which the human individual would like to remain, Gary Stix (2006) apprises in an article in *The Scientific American* entitled "Owning the Stuff of Life". We can read this textually: "There is a gene in your body's cells that plays a key role in early spinal cord development. It belongs to Harvard University. Another gene makes the protein that the hepatitis A virus uses to attach to cells; the U.S. Department of Health and Human Services holds the patent on that." Here unfolds the path that begins to mark off the patenting of human genetics.

The Progressive Worsening of Human Inequality

It has been about two and a half centuries (since 1754) but we are still suffering from the same disheartening evils: extreme social contrasts, exploitation of humanity by humans, environmental devastation, war and destruction. Undoubtedly, it is also a story of progress and achievements, evolution and success, but it remains a story of tragedies and domination, slavery and genocide. After recent major tragedies in the 20th Century, a great common law, the fruit or supreme bulwark of our humanity and dignity arose, after which all other rights might emerge, and that was, the Universal Declaration of Human Rights of 1948. It should not be forgotten that horrendous aberrations had to be committed in order that such an international bioethical declaration could finally be ratified by consensus in two hundred and thirty countries. Of course it is impossible to forget the atrocities that contributed to the unfolding of this process. But let us set that aside for a moment even while we acknowledge that such horrific events occurred in the recent history of what we call humanity and the civilized world. We can now argue that, if our question were to be: "What was the origin and the foundations of the *new* inequality between human beings, and was this authorized by natural law?" The response, as incredible as it might seem, would be the same that Rousseau presented to the Dijon Academy two-hundred and fifty years ago. In the first paragraph of his Second Discourse,[5] he harshly affirms—and this affirmation serves perfectly as an analogy for the commercial exploitation of the human genome and all other bioprospecting currently underway—that:

> The first man who, having enclosed a piece of ground, bethought himself of saying *This is mine*, and found people simple enough to believe him, was the real founder of civil society. From how many crimes, wars and murders, from how many horrors and misfortunes might not any one have saved mankind, by pulling up the stakes, or filling up the ditch, and crying to his fellows, "Beware of listening to this impostor; you are undone if you once forget that the fruits of the earth belong to us all, and the earth itself to nobody."
>
> 2005: 123

And so now, I similarly shout out loud and clear: to patent the human genome is a spurious work carried out by those who are detractors of their own humanity, aided and protected by unscrupulous scientists and greedy businessmen! And here I say: let us remember, my human brothers and sisters, while there is still time, that life belongs to all living beings and cannot be the exclusive property of anyone in particular! From this reasoning arises the following question: those early patents placed on human life, awarded in bits and pieces, are they not just like those first impostors, staking out their interests in the promising new fields of science and technology related to life itself? Analogously, are we not ourselves in a situation similar to the one in which the philosopher Rousseau found himself, who predicted albeit from a temporally bound enlightened perspective a bleak future for humanity? Moreover, what an uncomfortable and disconcerting impotence that we experience in the face of

---

5 *Discourse Upon the Origin and the Foundations of the Inequality Among Men* consists of two parts.

the development of these dynamics within contemporary techno-sciences that paradoxically move beyond our objective and intentional control, thus generating hypertely.[6]

Could the modern day counterparts of Rousseau posit a brighter future for humanity? Can we, the postmodern observers, if not completely retreating in the face of the techno-sciences, at least prove capable of managing them in a way that does not generate such great risks of harm to all of us? The initial indications seem to suggest that the answer is no, and here I regretfully must agree since there is no effective social control with respect to the ongoing developments and technological advances themselves. This development is something that happens *for itself*. It is clear that happens *for itself* means that a multiplicity of actors and elements are together aligned in the same sense of ingenuity and cleverness, where the human being continues to push up against the frontiers of the possible and the conceivable. The laboratory or laboratories armed moreover with extremely high financial backing for the concertation of all such factors operate under the protection of complex laws of industrial secrecy. It is in those very special places, the so-called laboratories, where human beings transcend their individual capabilities, extrapolate the foreseeable expectations, and feel perplexed in the face of such a potent and demiurgical power that is capable of producing an "Oasis of Technoscience." This was affirmed by Bruno Latour: "The structure of the laboratory exudes respect for the lego and correctly so. There do not exist many places under the sun where so many and such solid resources, are sedimented in such strata and capitalized on in such a grand scale" (2000: 90).

Where does all of the money come from? Who finances all of the mega-infrastructure of such a highly sophisticated platform of laboratory science? In the first place, there are the warlords—the USA followed at a considerable distance by the European Union, Japan, United Kingdom, and the rest of the developed countries. The participation of Brazil with respect to the wealthier countries would be comparable to residual fragments in the face of an enormous and monumental solid bloc. Directly wrapped up in these dynamic processes of development of the highest technologies is likewise the private sector which includes the vast majority of global entrepreneurs and volatile international capital. The largest financial groups and conglomerates have their own R&D operations and work under the preferred regime of secrecy and large figures.

---

6   Hypertely refers to the possibility that the development of something exceeds the ends for which it was designed and created. (Simondon 2007: 71).

On the other hand, the rest of us in society who know something about the state of the art of bio-nanotechnologies are familiar only with the tip of a gigantic iceberg which contains an ocean of R&D institutions and secret experimentation of six or seven times the size of that which is immediately observable on the surface. The fact is that these areas are key to the development of new platforms such as the converging technologies of NBIC.[7] The secrecy comes first, followed by the notion of what should be unveiled for the outside world and then undoubtedly commercialized. The gravitational point of these thoughts and sentiments of the post-modern individual is the commercialization of products, or more specifically, profit. And just as in the case of the ancient man that Rousseau shows us, the question emerges: How to stop events or experiments from occurring when their very existence is barely known of? If then the real dynamic is imposed by the market, knowing this in advance allows us to not be deceived for one moment: the highly profitable activities will continue. The problem is that there are no established limits by which to guide the dynamic and commercial activities of the techno-scientific developments that products and inventions create. What occurs for techno-sciences also occurs as singular, emergent, random and fragmentary phenomena within that same recursive environment of unregulated activity aimed at the realization of profit. Nevertheless, there needs to be a way of remaining alert to what all of these things signify in anthroposocial terms for humanity, since societies generate technologies that inevitably transform them as has been already pointed out. We can add to this that the intense velocity of such phenomena, i.e., the rapidity with which events move from the hypothetical field to the field of reality, and the concentration of wealth that this will represent for those who control these patents,[8] almost certainly poses the risk of serious crises for humanity in the political and bioethical spheres just as it will certainly lead to increased inequalities among human beings. The logic of the industrial patent is very simple: those who hold it can commercially exploit it while the rest of the world must pay a premium to be able to consume it. Once more, the geographical distribution of this exploitation and its human agents are similar to those who were at the top in the times of Rousseau,

---

7   Converging Technologies are stimulated by advances in four core fields: Nanotechnology, Biotechnology, Information technology, and new technologies based in Cognitive science (NBIC).

8   Industrial patents are legally framed in such a manner as to allow their owners to commercially explore a technology that has been developed over a specified period of time, while the remainder of the planet that wishes to use or industrially apply the patented object is obligated to pay royalties to the owner.

with the singular exception of the USA which did not yet exist as an independent power. But in the present day, it is exactly the opposite as Latour points out: "Half of all technoscience is under the dominion of the United States. The rest of the developed countries are working on smaller portions of science" (2000: 161). From the year 2000 to the present day, as Latour suggests, this hegemonic scenario is becoming ever more pronounced and this worsening socio-political-cultural conjuncture is becoming ever more evident, with 70% of the resources involved destined for the Pentagon and the US Department of Defence for armaments and military technologies that can support armed conflicts everywhere. What is truly horrifying is also what generates the desired and celebrated profits we have already alluded to. It is important to be clear, moreover, that we later become oriented towards consuming the results of war industry developments, once converted into less offensive goods that are marketable and profitable. The Internet is the classical example, just as modern aviation was in an earlier era. The products that follow the same trajectory are extremely diverse, ranging from Velcro to space vessels, including products such as cellular telephones, armaments, and toys. According to Latour: "[...] to a very fortunate few are reserved the invention, discussion and negotiation of all this while the remaining billions of inhabitants remain without any other recourse but to borrow the use of these products for their black boxes, or else remain in total ignorance" (2000: 161). Additionally Latour affirms that "Vomited out by some research centres and laboratories, new objects and beliefs emerge, freely floating through minds and hands, populating the world with replicas" (2000: 128).

A Grave Threat to Democracy

As Winston Churchill aptly summarized on one occasion, "Democracy is the worst form of government, except for all the others." This is something that in effect we can assent to. It is the best and most developed political form developed over time to viably manage everyday social life. Born more than two thousand years ago in Ancient Greece in an extraordinary period of humanity known as the Age of Pericles,[9] democracy continues to be practiced to the

---

9   Pericles (495–429 BC) was a political figure born in Athens who had as his teachers the philosophers Anaxágoras and Zenón de Elea. He was heavily involved in Athenian politics and known for being rational and optimistic, a great orator, and above all, for his strategic method of introducing valuable democratic forms, most of which were accomplished by

present day as a form of political regime which as we know allows for a relative degree of social participation via the election of representatives in making decisions for the larger social collectivity. The term *democracy* with its origin in Ancient Greece literally signifies government by the people. As we know, democracy is a political regime in which on the one hand, all citizens are eligible to participate, and on the other hand, they likewise participate in the election of their representatives and in the presentation of proposals, their development and the creation of laws, in such a manner that it permits political governance of a people and of a country, consummating its political self-determination. The democratic system principally contrasts with other forms of the political exercise of power in which power is invested in the rule of a single person, for example, in the form of a monarchy, tyranny or dictatorship, or by the rule of a small group of people as in the case of oligarchies. The democratic system, amidst its numerous unquestionable virtues, makes it possible for a society to control its governing bodies and political representatives, including removing governing representatives from power without (and this is important) the necessity of a revolution. Moreover, as Edgar Morin puts it:

> Democracy is a historical achievement of social complexity. As we have indicated, it instates rights and liberties for individuals, as well as elections that assure the control of those that control by those that are controlled, in a climate of respect for the plurality of ideas and opinions, the expression of antagonisms and their regulation that impedes their expression in violence. Democratic complexity, when it is well rooted in the history of a society, is a metastable system that has the quality of surviving internal conflicts, innovations and unexpected occurrences.
>
> 2006: 165–166

In summary, democracy permits the discussion and deliberation over issues relative to collective interests with greater horizontality and popular participation. The same society that practices it can opt for alternatives, paths and decisions that it takes. The friction of contrary ideological positions, the broad discussion of political measures and proposals, democratic votes, all together

---

reforming Athenian legislatures in favour of elevating the power of the majority in opposition to the power of oligarchic interests. The Peloponnesian War and later a plague that afflicted Athens weakened and ultimately took his life.

generates ideal conditions so that the common interests can be contemplated with greater effectiveness. Echoing our thoughts and those of Edgar Morin, Andrew Feenberg writes:

> Democracy demands public discussion of ideals in the context of the free circulation of propaganda, of the influence posed by business interests and of technologically deterministic ideologies.
>
> 2010: 331

This technocratic power that Feenberg speaks of has in an accelerating manner fundamentally degraded the seminal values that we hold and use in order to give meaning to life and human relations, including the interpersonal relations of equality and solidarity, the formation of interacting social groups, and the broadest interests of humanity. This can be seen in the case of genetic patenting, threatening in this way democracy itself through the commercialization of life, transforming into the end what should morally be only the means of achieving ends, transforming into commercial goods its basic and structural elements, freely commercializing it like any other commodity, in such a manner that life itself turns into a mere product. If on the one hand, genetics can bring precise benefits such as aiding in the prevention or treatment of specific illnesses; on the hand, we become thrust into a kind of complicated bioethical dilemma where the commercial interests of some are superimposed upon the collective interest of the vast majority of citizens. It is fundamentally important to observe the importance of this fact, given that we speak here not of social segments but of the human species as a whole. Francis Fukuyama adds:

> Human nature moulds and limits the possible kinds of political regimes, in a way that a technology that is sufficiently powerful to transform what we are, can possibly have malignant consequences for liberal democracy and the nature of politics itself. [...] But one of the reasons why I am not so optimistic is that biotechnology, in contraposition with many scientific advances, mixes obvious benefits with harm in an unsewn package.
>
> 2003: 21

Democracy is a civilizing and politically recent creation if we consider that the long history of homo sapiens on the planet was achieved only with considerable effort, by way of bloody and dark periods of human history. Humankind itself as a socially organized species now sees its achievement dramatically threatened. To patent portions of the human genome, patents exclusively aimed for commercial exploitation of the technology, is something that contravenes the sense of fraternity and equality that diametrically and socially unites us. Edgar Morin also informs us that "the increasing dispossession of citizens to access over control and from reflection on scientific or technical knowledge concerning the life of everyone is leading to a decline of democracy precisely at the point where democracy is rooted" (2003: 217). Elsewhere, Morin affirms that "the continuation of the current techno-scientific process which to all others remains in the blind also escapes the conscience and will of the scientists themselves, leading to a sharp reversal of democracy" (Morin 2006:170). Michio Kaku has written "in a democracy, what most matters is lucid debate by an enlightened electorate" (2001: 305). In agreement with those arguments, Hugh Lacey also informs us that:

> At the present moment, the practices of controlling nature are in the hands of neoliberalism and, in this manner, serve certain values and not others. They serve individualism rather than solidarity; to property and profit rather than social goods; for the market and not the welfare of all people; for raw utility and not the strengthening of the plurality of values; for individual freedom and economic efficiency rather than for human liberation; for the interests of the rich and not the rights of the poor; for formal democracy rather than participatory democracy; and to civil and political rights without any dialectical relationship with social, economic and cultural rights.
>
> LACEY 1998: 32

Lacey's tone is a bit sceptical and little hopeful but far from being an exaggeration. It reflects that the sheer economic tyranny of the socio-technical reality we experience today is the new barbarism invented by humanity. In the name of efficiency and principally profit, almost everything is allowed, and techno-science, in this context, is the driving force behind development itself. It is important to be clear that the less fortunate and privileged are considered necessary only so that capitalist business can keep profiting. It is this mass of consumers, hungry for technical objects, that moves and justifies the entire chain of production, and genetics in this regard is no exception. Once inside the business and commercial scheme, little or nothing can be done to protect and safeguard life. The intended benefits, as we know, arrive latest to the neediest, if they in fact arrive at all. The organizational scheme does not permit very much manoeuvring. All that precedes her path toward higher gain is, may be, and usually is discarded. Noam Chomsky writes that:

Fascism is a political term which is not applicable in the strict sense to enterprises. Nevertheless, we can observe that power always flows from above: from executive directives to the managers and directors, and ultimately, to the venders, secretaries, etc. There is no flow of power or planning in the opposite direction. The ultimate power resides in the hands of investors, owners, bankers and the rest. The people can complain or make suggestions, but the same is true in a slave society. Those who are neither owners nor investors have little to say. They can opt to sell their work to an enterprise or buy the goods and services that it produces, or look for insertion in the chain of command, but no more. The control over the enterprise is ever more reduced.

[...] Accidentally, much of what [the enterprises] do has beneficial consequences for the larger population. The same can be said about the government at whatever level. Of course, they are not interested in a better life for the workers, but rather higher profits and participation in the market. It is no secret and they are things that people should learn in third grade. Private enterprises seek to obtain the highest profits, power, participation in the market and control over the state. At times, its decisions favour others but only by coincidence.

1997: 14–15

This ought not to be the only role of business enterprises. Private companies are always looking for higher returns and profitability as their goals, and genetic technologies are no exception to that general rule. Technology always means power, especially over those who do not hold it or those who have inferior technologies, and also over those who necessarily need to pay in order to acquire it. In the final instance, that kind of power, so widespread and consolidated, will lead to private control of human biology, meaning that companies and groups will position themselves for the right of possession and commercial exploitation of our very genetic baggage.

As Renato Dagnino suggests in returning to the spirited debates surrounding epistemological issues and the bio-politics regarding the techno-sciences:

According to that school of thought [the Frankfurt School], technology is one of the most important resources of power exerted over modern societies. Decisions that affect our daily, political democracy have been obscured by the enormous power wielded by specialists of technical systems: corporate and military leaders, and groups of professionals such as physicists and engineers' associations. They exercise much more control over the patterns of urban growth, configurations of housing, transportation systems and the available selection of innovations and, in general, over our experience as employees, patients and consumers, than all of the government institutions of our society.

2008: 114

What moves techno-scientific industry and commerce is the stubborn search for results and applications, utilities and operationalities, all relentlessly predicated upon the "soul of business" in capitalism that it is going to sooner or later revert to more sales and business, increased profitability and profits. Not to mention that the whole process of technification takes place extremely quickly, to a certain extent thoughtlessly as the times involved literally run over any possibility of the necessary epistemological reflection. In short, the process does not leave enough time to reflect on the larger significance of these socio-technological dynamics that are forcibly being imposed. Our technical capacity to carry out projects and create new technologies far transcends our ability to categorize and conceptualize these same developments. David Le Breton asserted "We witness today an acceleration of social processes without a corresponding adaptation of culture. It is possible to speak of a divorce between the social experience of human agency and its capacity for symbolic integration" (Le Breton 1995: 14). Capacity for symbolic integration in this context signifies being capable of understanding the meaning and the consequences of what we are doing in the social sphere. Moreover, if we begin with the assertive and presumably correct assertion that techno-sciences are not nor can they ever be neutral, given that they flower out of the investments of global capital, we perceive that complex problems remain with respect to the conflicts of interest between means and ends. As Max Horkheimer recognised: "The social genesis of problems, the real situations in which science and the goals pursued in its application, become considered themselves external to the process" (1975: 163). To this, we can affirm along with Michio Kaku that "in a democracy, only debate engaged in by the citizens can lead to mature decisions about such a technology so powerful that it allows us to dream about controlling life itself" (2001: 284).

What we want to highlight is that the essence of the idea of democracy is found in the concept of equality of rights and duties among persons, and of reciprocal likeness among all human beings. However, genetic patenting extends an economic and entrepreneurial imperative that already dominates other fields and ignores essential values, making some people less privileged and vulnerable to control and exploitation by others who are more affluent and profit hungry, all of which goes against the basic grain of democracy, equality and human dignity. It suggests that a human being can become transformed into a means for any purpose which is not its own, that is to say, beyond its very existence. In other words, third persons can come to wield the key to our biomolecular constitution, or at least significant parts of it, and this will most assuredly open the door to a new era of atrocities and barbarism for humanity.

Prospects and Horizons for the Techno-Sciences

In view of our various critical and theoretical reflections, including those of thinkers from widely varying fields of human knowledge, we can re-engage our main question: "What is the origin and foundations of the new inequality between men, and is it authorized by natural law?" In this regard, we can clearly see that over the course of history, these early inequalities became deepened and proliferated in accordance to the degrading structures of nascent capitalism and the underlying laws of the market that it fuelled, revealing the ghostly presence of profits as a complicating and aggravating factor. Thus in the present day, the ability to access the genetic background of the world's peoples signifies the means to control, manipulate, regulate, commercialise, and, why not say it, discriminate since the underlying logic of the game in vogue is exploratory efficiency for nothing more than commercial gain.

If we grant that we sell our time in wage labour in exchange for money, wilfully selling our labour power, what then are the limits to curtail this form of mercantilist reasoning with respect to such things as stem cells and genomics? Would it not follow, then, to evaluate candidates for a systems analyst position in a company according to the medical records that contain information about their likely longevity and possible susceptibilities to specific diseases, or other data related to their health? What would be the moral obstacles to be considered, once life could be patented and freely bought and sold? The same regime of jobs currently in place within our culture would become transformed into a slave reality rather than one of productive remuneration, save for rare exceptions. The vast majority would simply sell themselves in exchange for small and insignificant amounts of money. And there would be few possibilities of upward mobility for entire groups who have to spend their earnings on their subsistence, leaving practically nothing left over from their strenuous work lives but to grow old and die, with little if anything to pass on to their descendants. This is what is occurring in the present day, just as it did in the time of Rousseau:

> Such was, or may well have been, the origin of society and law, which bound new fetters on the poor, and gave new powers to the rich; which

irretrievably destroyed natural liberty, eternally fixed the law of property and inequality, converted clever usurpation into unalterable right, and, for the advantage of a few ambitious individuals, subjected all mankind to perpetual labour, slavery and wretchedness.

ROUSSEAU 2005: 142–143

It is impressive to reflect upon the relevance of this discourse by Rousseau in the present day. In this sense: would it be appropriate to ask ourselves if we too would be the willing victims of a brutal and irrevocable usurpation of life, where our own genetic background would be a sellable good, and where an unethical and greedy few could become the norm, crystallizing into an established and irreversible kind of society? Here I leave to the wise reader the mission to reflect upon this.

We still have the matter of our question "[...] and is this new inequality authorized by the laws of nature?" The answer is clearly that it is not. We cannot blame nature for social ills that are solely and exclusively human in origin. Such injustices have no other source than the reciprocal social interactions and commercial logic that produces very impressive effects which are neither reciprocal nor equivalent, in regard to the distribution of the wealth generated by those very same predominantly unequal relations, nor in regard to the appropriation of the technologies used to generate this wealth by those populations that create and consume them. We return again to Rousseau, our master of ceremonies in this critique of post-modern technologies, to realize that the social inequality we are referring to is an evil that has been with us for a long time:

Do you not know that numbers of your fellow-creatures are starving, for want of what you have too much of? You ought to have had the express and universal consent of mankind, before appropriating more of the common subsistence than you needed for your own maintenance.

ROUSSEAU 2005: 141

As a predominantly technological civilization, we also have, in this singular momentum, a truly infinite number of open possibilities before us due to the techno-scientific conjuncture and convergences ever at the service of exploitation and of those imperialist actors always behind them. All of this power, when added to the critical incapacities of societies that in general look to science and technology as essentially good things, sees as beneficial that cultures encourage their development, generating an atmosphere of unlimited scientific promiscuity that allows for abuses or delinquency. Techno-science is not necessarily a bad or good thing, however, it is easy to see that it serves the master that it is offered to, and is particularly vulnerable and susceptible to the power of capital and industrial market capitalism. One of the eminent risks, in this sense, is that the market will devour everything, i.e., prove able to overcome everything else in terms of operativity, prevalence and especially as a truth (and/or belief) as Carafe and Berlinguer point out: "One of the risks is that the laws of the market subvert any and all moral principles" (1996: 43). We also see that with regard to our critique of technologies, the arguments of Ernest Gabor (a lab partner of Einstein) and the philosopher Friedrich Nietzsche, as well as the trans-disciplinary convergence of both in the criticism of existing cultures, in what we now call post-human and indiscriminate techno-scientific progress. Gabor thus advocates the theoretical postulation known as the Law of Gabor: "Everything that can be done, it will necessarily be done." Nietzsche for his part considers and indirectly confirms this notion:

> [...] we have a still undiscovered country before us, the boundaries of which no one has yet seen, a beyond to all countries and corners of the ideal known hitherto, a world so over-rich in the beautiful, the strange, the questionable, the frightful, and the divine, that our curiosity as well as our thirst for possession thereof, have got out of hand alas! That nothing will now any longer satisfy us! How could we still be content with the man of the present day after such peeps, and with such a craving in our conscience and consciousness? What a pity; but it is unavoidable that we should look on the worthiest aims and hopes of the man of the present day [...] the ideal of a spirit who plays naively (that is to say involuntarily and from overflowing abundance and power) with everything that has hitherto been called holy, good, inviolable, divine; to whom the loftiest conception which the people have reasonably made their measure of value, would already imply danger, ruin, abasement, or at least relaxation, blindness, or temporary self-forgetfulness; the ideal of a humanly superhuman welfare and benevolence.
>
> NIETZSCHE 1974: 231

In this steep assent in which we risked the celebrated creation of a set of complex and sophisticated technological practices, allowing us an infinite number of applications in terms of bioprospecting and control, it is clear that we are facing a war. It is a war waged by increasingly greater and more sophisticated knowledge and techniques, which allow their holders nothing less than reinventing matter, life and the human species itself, and if that were not enough, to make a lot of money in the process. The improprieties do not stop there: we can also create new forms of life that never existed on Earth, and can even patent these living creations because, since the last decade, such discoveries of partial or even complete human DNA sequences can be patented freely. As Carafe and Berlinguer affirm:

> That procedure to plant the flag on our domestic territory, to insist on possession, has many similarities with the property rights claimed by the conquistadors of past centuries who declared themselves owners upon discovery and colonization of the inhabitants of territories. At that time, it was on behalf of the King of Spain or the Queen of England, or one or another religion. Today, it is in the name of science, but always with the thirst for profits.
>
> 1996: 61

Closely related is the polemical issue of human cloning which also presents profound bioethical dilemmas. Now after the famous cloned sheep Dolly, and since that cloning technique is virtually identical to that required for a human, a complete human clone could appear at any time. However, what is not usually reported is that in order to clone this poor sheep, there were 270 failed attempts. It is worth noting that in each of these 270 failed attempts, animal life was created and ended up deformed, stunted, degenerated with flaws and serious congenital problems, and often ended up literally in the trash, a practice that certainly symbolizes the undignified exploitation and instrumentalization of other living creatures who are totally defenceless and at the mercy of such cruel atrocities. These creatures certainly underwent all kinds of torture and unimaginable suffering before a thing like Dolly could have been made to appear as something "positive" and presentable as a product of scientific research. However, despite this horrific case, there are still many experts actively involved with such projects. To create human clones for the purpose of exploitation and utilisation of their separate vital parts is another repugnant development that is arising on humanity's horizon, hyper-empowered with the advent of the techno-sciences. Volnei Garrafa and Giovanni Berlinguer indicate:

> Naturally, we immediately were curious to know if the clones are people and, therefore, are entitled to the respect due human beings for equal protection. Dr. Roderic Gorney suggested that this problem could be overcome by keeping the clone [or clones] in an artificially induced unconscious storage so that the mind and human personality does not develop.
>
> 1996: 159–160

In the face of all that has been discussed this year, the bioethical doctors Garrafa and Berlinguer indicated with high certainty that: "Mine, yours, ours, the autonomy of everyone is violated by the declaration of property presented by an individual who belongs to our same species" (1996: 138). Jean-Jacques Rousseau in his Discourse was filled with disconcerted reflections in its final pages, referring to events of the past, but that even still apply perfectly well to the present day:

> [...] he will explain how the soul and the passions of men insensibly change their very nature; why our wants and pleasures in the end seek new objects; and why, the original man having vanished by degrees, society offers to us only an assembly of artificial men and fictitious passions, which are the work of all these new relations, and without any real foundation in nature.
>
> 2005: 241

Conclusion

A difficult and at the same time promising truth which we carry into the new millennium, if we dare say it, is exactly to have to live with the idea that all things are possible, at least theoretically, in terms of techno-sciences. That truth can be expressed twice over for human civilization in dynamic and systemic terms. In light of this fact we may know how to administer the awesome power that is opening up before us, and we may perhaps manage the evil even if we continue to move forward in terms of technological development without pondering and evaluating the consequences of these same processes. Or, perhaps well beyond both possibilities, we may simply succumb to the "contingent factors" of "disorder" and the "unpredictable", as Luhmann (1980) puts it. So, this quasi-divine power of interference and technicist control that we construct ourselves with our own talent and wit, at once promises us a thousand and one miracles in technical terms, but it can equally be harmful if used improperly and indiscriminately. There is no clear way to limit the process of technological progress, but certainly we have to develop the critical capacity to anticipate, prevent, and especially to manage the consequences of our own powers. Just as Rousseau, we no longer have any doubts: "It is sufficient that I have proved that this is not by any means the original state of man, but that it is merely the spirit of society, and the inequality which society produces, that thus transform and alter all our natural inclinations. [...] It follows from this survey that, as there is hardly any inequality in the state of nature, all the inequality which now prevails owes its strength and growth to the development of our faculties and the advance of the human mind, and becomes at last permanent and legitimate by the establishment of property and laws" (Rousseau 2005: 164).

We can affirm then, in accordance with Jean-Jacques Rousseau, that this humanity that creates innovative technicist practices, will be the same that will have to find ways of controlling its own exacerbated, deregulated and technicist lack of control, especially in the delicate and complex biological fields. This is where it is most essential to retain full respect for the bioethical precepts of every free and autonomous human being, preferably, in the democratic exercise of inalienable individual freedom. It is likewise imperative to uphold the set of Universal Human Rights as a framework which, beyond whose boundaries, irresponsible entrepreneurs, alienated scientists, and even arrogant and belligerent nations, would not dare to cross lest they risk the severest punishment in accordance to the dictates of international law collectively established around the globe. Any such techno-scientific incursion that so transgresses the most basic values and structures of present human civilization is surely unworthy to place into practical operation.

CHAPTER 2

Unemployment, Inequality and Technological Development

Víctor Manuel Figueroa Sepúlveda

Problems of unemployment and social inequality have become the dominant concerns of political economy in recent times. The true meaning of these phenomena and their causes are the object of colourful theoretical and statistical debates. In this work, we intend to explore these issues given their evident importance for understanding the present day reality. Special attention will be dedicated to their relationship with technological development, with a focus on the case of the United States, the most technologically developed and most unequal country. We will conclude with a brief discussion on the significance of the current changes in the political organization of society.

Data

It is estimated that total world unemployment currently has reached the level of 201 million persons, some 30 million more than in 2007 and it has been projected that in order to address this situation, it would be necessary to create around 470 million jobs by 2030. This condition of the world economy and the challenges it represents involves all types of countries, although in an unequal manner. Table 2.1 provides the rates of unemployment for some of the developed countries.

In the case of Europe, the unemployment rates in France, Italy and the United Kingdom are alarming and there is no real tendency presently evident towards a real recuperation of employment. On the contrary, the situation has markedly worsened since the year 2007. Only in Germany has unemployment been reduced even if simply stabilized at elevated levels.

The unemployment rate for the developed countries in 2014 was estimated at 8.6%, precisely where for many years these same countries were considered to have the greatest capacity for generating employment. There can be little doubt that profound changes are taking place in the unfolding of capitalism and these changes will need to be understood in light of development itself.

TABLE 2.1   *Rate of unemployment (% of total)*

	2007	2012	2013	2014
United States	4.7	8.2	7.5	7.2
Japan	3.9	4.3	4.1	4.0
France	8.0	9.9	10.5	10.9
Germany	8.6	5.4	5.3	5.3
Italy	6.1	10.7	12.1	12.6
United Kingdom	5.4	8.0	7.5	7.3

SOURCE: ORGANIZACIÓN INTERNACIONAL DEL TRABAJO (2014)

TABLE 2.2   *Rate of growth in wages for selected developed countries,*
             2007–2013

United States	1.4
Japan	−1.3
France	2.3
Germany	2.7
Italy	−5.7
United Kingdom	−7.1

SOURCE: ORGANIZACIÓN INTERNACIONAL DEL TRABAJO (2014)

Wages under conditions of mass unemployment have little or no hope of a significant recovery. Table 2.2 provides the growth rates over the period of 2007 to 2013. Unemployment likewise contributes to a widening gap between the rate of productivity and the rate of growth of average wages (Table 2.3). This continues a trend of annual growth rates across developed countries (Table 2.4) that began over the last two decades of the Twentieth Century.

The portion of wages in income has fallen in all the countries shown above. The composition of income has evolved contrary to wages and for that reason contrary to wage goods in relative terms.

In the United States, the average real salary per hour grew 76% between 1947 and 1972. Since then until 2011, it grew only 9%. The bulk of this increase was absorbed by executives, particularly those occupying the highest positions.

With respect to unemployment in the United States, the following points need to be made:

TABLE 2.3   *Index of productivity and average salaries of developed countries, 1999–2011*

Year	Productivity	Average wages
1999	100.0	100.0
2001	103.6	101.5
2003	106.5	103.2
2005	109.9	103.2
2007	112.8	105.3
2009	110.7	105.8
2011	114.6	105.9

SOURCE: ORGANIZACIÓN INTERNACIONAL DEL TRABAJO (2012)

TABLE 2.4   *Growth in wages of developed countries, 2007–2013 (%)*

2007	2008	2009	2010	2011	2012	2013
1.0	−0.3	0.8	0.6	−0.5	0.1	0.2

SOURCE: ORGANIZACIÓN INTERNACIONAL DEL TRABAJO (2014)

- The prevalence of unemployment is now higher. It went from a lapse of 3–4 months in 2003 to an average of 6 months in 2012.
- The number of long term unemployed individuals, i.e., those out of work for more than 27 months, rose from 2.6 million in March 2014 to 3.6 million in March 2015.
- Those who had not been seeking work in the last four weeks, despite the fact that they were out of work, are not counted. In March 2015, there were 2.1 million unemployed persons no longer actively seeking work during the previous four weeks.
- Individuals are counted as employed even if they carry out part time labour activities regardless of their desire for full time work. Officially figures calculate that the number of workers that fall into this underemployed category reached 6.7 million individuals in 2014. This amounts to a significant distortion in employment figures. At that time Paul Krugman had been working alongside of government efforts to construct more accurate measurements. He noted that it could be seen that unemployment in 2011 reached an effective rate of 15% (Krugman 2012).

Over the long term, official US figures, without taking into account the additional observations noted above, show the following historical pattern in unemployment as indicated in Table 2.5. Since the decade of the 1970s, US unemployment had stabilized at about the level considered to signify "full employment" (around 4%). The years of growth between 1995 and 2000 and later between 2003 and 2007, propelled by speculative bubbles, had a positive effect on unemployment rates as shown in the table. Between 1980 and 2012, employment by sector grew in services (architecture, engineering, computer design systems, management consultancies, technicians, health, and small business) from 39.8% to 48.7% and fell in agriculture from 2% to 0.9% as well as in industry from 17.3% to 10.1%.

The projections for 2022 are not very promising. The average rate of growth in employment that was 1.2% between 1992 and 2002, fell to 0.7% between 2002 and 2012 and is estimated to fall further to 0.5% between 2012 and 2022. It is expected that employment will grow in health services and social welfare services at a 2.6% annual rate, absorbing one-third of the projected growth. For its part, construction is forecast to contribute a 2.6% growth in employment. Employment will continue falling in the manufacturing sector, federal government services, the forestry and agricultural sectors, and in the fishing and livestock industries. On balance, the tendency will continue in which each decade will see a lesser quantity of employment. Between 1992 and 2002, employment grew by 13.1% while between 2002–2012, it grew by 7% and the projected growth for the period between 2012 and 2022 will grow only 5.5% (Bureau of Labor Statistics 2014). Inequality that goes hand in hand with unemployment along with the social protests that it has provoked, has led to a growing body of research. One influential example is the work of Thomas Piketty (2014) that has rekindled the discussion concerning social inequality and work. His tortuous and stringent work with the statistical data has culminated in his widely acclaimed *Capital in the Twenty-First Century*, which provides a welcome contribution to the pragmatic atmosphere that prevails in these times of neoliberalism. In spite of a certain neoclassical blind spot, Piketty does not identify his work with any particular theoretical tradition and in fact works hard to keep his study at the margin of any implication this has for society.

It has been said that the results of Piketty's study have done little more than confirm the theoretical positions of Marx. But in reality, these two positions are very far apart in terms of their epistemological perspectives as well as their postulates about the path that capitalism is taking. Piketty's concept of capital that is identified with wealth rather than the social relations of exploitation is a palpable expression of their theoretical distance. It is doubtful that Marx

TABLE 2.5   *Evolution of unemployment in the United States (%)*

1950–1959	1960–1969	1970–1979	1980–1989	1990–1999	2000–2009	2010–2014
4.6	4.9	5.6	7.2	5.3	5.0	8.3

SOURCE: BUREAU OF LABOR STATISTICS (2015)

would have chosen to focus study of the capitalist economy on the relations of distribution as Piketty insisted. Or that the problems of inequality could be resolved or even mollified within the framework of this form of society by managing taxation or inheritances, even if this was an issue that Piketty did not feel particularly strongly about. The author didn't wish to get wrapped up in theoretical controversies, not even within neoclassical circles, thus providing a new focus on inequality that according to the data will continue growing in spite of earlier approaches to the issue.

In any case, *Capital in the Twenty-First Century* deserves a separate and dedicated analysis. The enormous richness of the information offered (which even includes a virtual link to statistical appendixes), the complexities of its mathematical formulas, its impact in economic history, etc., in addition to the policy concerns it has generated, all point to the book being a work worthy of careful study.

The realities that it describes amount to unquestionable features of the present capitalist society. The consensus on inequality is practically universal even if technical discrepancies abound regarding the phenomenon. Studies in general coincide on the following points: a) after a period (1928–1978) in which inequality diminished, the income of the wealthiest sectors rapidly grew at higher rates. The 10% of the population that in 1980 controlled 35% of the wealth came to wield 50% by 2008; b) the income of the remaining 90% of the population has remained practically stationary since then; c) the differences within the highest layer of the wealthy has deepened, according to the differences of wealth that they control and the increase of income that has been obtained among the diverse segments. The 0.1% of the US population by 2013 managed to control around 23% of the national wealth whereas in 1979, they controlled only 7% (Sáez and Zucman 2014); d) the so-called middle classes have been disintegrating, possibly to a large extent in accordance with the educational levels that its members now possess; and e) poverty affects around 49.6 million people. A similar number of people lack health care coverage.

Causes

In general, it is accepted that unemployment just like inequality results from a variety of causes even as specific studies place their emphasis on one or another variable. The following variables are illustrative examples among others:

- Low rates of growth in the context of a long term tendency towards decline. Not only does unemployment grow but so too does the competition for positions of work, as well as the size of those sectors with precarious wages.
- The growth of financial capital that has allowed income to shift from labour to capital has also given rise to an enormous unregulated financial system that makes possible the increase of compensation to the financial sector, thereby weakening the rest of the economy and encouraging destructive practices in the manufacturing sector, stimulating greater indebtedness (Gordon 2014).
- The decline in education. Higher education no longer enjoys its mass character. Young adults and children living in poverty now lack the conditions for beginning their studies and their graduation rates have not kept pace with those from wealthier sectors.
- The decline in union organization, the lack of a minimum wage policy and unemployment insurance.
- Globalization and its tendency to diminish the importance of the domestic market.

The effects of technological change upon employment have been widely recognised and have lately captured the attention of economists. A notion utilized by Keynes in 1930 is that of "technological unemployment" and this has been dusted off and put back into use in order to legitimatize the concerns of the present. What Keynes said back then was the following:

> For the moment the very rapidity of these changes (those promoted by technological progress) is hurting us and bringing difficult problems to solve (...) Those countries are suffering relatively which are not in the vanguard of progress. We are being afflicted with a new disease of which some readers may not yet have heard the name, but of which they will hear a great deal in the years to come—namely, technological unemployment. This means unemployment due to our discovery of means of economising the use of labour outrunning the pace at which we can find new uses for labour.

Nevertheless, Keynes immediately adds the following:

> But this is only a temporary phase of maladjustment (...) I would predict that the standard of life in progressive countries one hundred years hence will be between four and eight times as high as it is today. There would be nothing surprising in this even in the light of our present knowledge. It would not be foolish to contemplate the possibility of a far greater progress still.
>
> KEYNES 1930: 363

It is certain that Keynes' observation concerning the conflict between the rhythm of technological change and the capacity for finding new sources of employment at the same velocity is felt more present today than ever. His prediction about the general well-being did not pan out, at least not for large masses of the population. Keynes saw in this conflict a temporary maladjustment, and this seemed to be confirmed with the arrival of the "glory years" but in reality, the conflict has become accentuated in recent decades. What is important to clarify is if even still, and also in the very long term, the reality of the most recent decades could be thought of as simply a period of maladjustment, one that could be resolved with state interventions. The most relevant question for the present is if in truth capitalist society will be capable of providing a response to the demand for employment. Everything seems to indicate that the answer is no.

The changes that have taken place beginning in the 1970s and 1980s have placed into evidence the long term relationship between accumulation and employment that Marx had anticipated. Growing surpluses of population have erupted in the scenario and the role played by technological progress has become ever more evident with each passing day. The work of Jeremy Rifkin (1995) *The End of Work* dramatically increased the attention being paid to this issue. Although he thought that among the causes of unemployment, it was necessary to take into account other factors such as globalization, he blamed technology as the principle and unceasing causal factor. But he did not treat this phenomenon as inevitable. He already thought at that time that the expansion of a new sector, the "social sector" could overcome the incapacities of the public and private sector. More recently, he has resumed his attempts to find a solution with more elaborate and complex propositions.

Nevertheless, there is a persistent scepticism about the negative relationship between productivity and employment. There is resistance to abandoning the positive sensations that these two variables have left recorded over an ever more distant historical record, as well as how it might have been superseded by

something really new over the last decades. Andrew McAfee has called attention to the enormous gap that has been opened since 1972 to date between
manufacturing and employment in this sector in the United States. In the same
manner he could identify in the data of the specialized statistical information
agencies, the futility shown by hesitancy in resistance. In observing the evolution of this gap, he adds:

> That really seems to me technological unemployment, especially when
> unemployment in the manufacturing sector has also been declining in
> Germany, Japan, China and around the world. When this is the case, it
> means that the changes in employment are not due to the displacement
> of work around the world in search of cheaper human labour; rather, it is
> due to the fact that machine labour is starting to become a least as capable
> as and cheaper than human labour.
>
> MCAFEE 2013: 1

McAfee sees nothing particularly problematic with all that. In effect, manufacturing employed 18 million workers in 1970 while by 2012 it had shrunk to only
12 million. One of the sectors most affected has been that of durable goods that
fell from 11 million workers to only 7.7 million over the same period.

If unemployment finds in technology an important causal factor, it is prob-
able that the same was happening with inequality and that one could begin
to speak of "technological inequality." This issue seems to be less interesting.
In the heart of capitalist society, inequality in technological capacities is an
intrinsic vocation that is carried out through a constant search for methods
that can allow producers to pass ahead of others with the aim of acquiring
an extraordinary profit. The difference rests in the objective: to arrive in first
place in the income rankings. Within labour, inequality has more to do with
the skills of each worker and with the conditions that allow the labourer to
apply his or her knowledge and skills. Technology in that sense can become
transformed depending on how it is conjoined into either an ally or an enemy
in the struggle for eking out a living. This inequality can effectively assist in the
comprehension of the present and foreseeable future. From a very abstract perspective, each historical epoch of labour had to define itself beginning with its
relationship to the means of production. The same is valid in the context of the
history of capitalism which places the focus on this relationship at this historical moment, and could well address the need for advancing towards a correct
reading of the present.

Piketty maintains that inequality takes place as a result of when the annual
rate of return on wealth is greater than the rate of economic growth. We have

not found until now an explanation in his work for this state of things, but his statistical information supports this assertion. A large number of authors maintain that the decisive cause of inequality is technological progress. In general, they pursue the idea that technological change favours the most educated workers in terms of income while at the same time increases the demand for them. This results in a constant and significant reordering in the distribution of wages.

Two categories of technologies have been detected according to these effects on employment. A) There are those that displace labour, such as robotics, digitally controlled machinery, computerised inventories, and automated transcription. These technologies effectively displace routine labour activities. B) Then there are those that require abstract reasoning and more informed labourers, such that this elevates their value to companies that demand highly skilled labour. These include those who carry out visualization of data, analysis, high velocity communications, and rapid design of prototypes.

The story of technological inequality goes more or less as follows. Those who are displaced and manage to update their skills in accordance with the demand for those skills can experience a leap upwards in their income. Those others who fail to continue updating their skills are destined to swell the ranks of the unemployed or will have assumed less demanding tasks in terms of requisite skills than previously exercised, the latter of which implies a corresponding reduction in income. In this way, whole groups of workers are dissolving through growing inequality gaps. In the lower-middle income stratum, new agglomerations are being formed in the labour market. Take for example the case of when routine tasks cannot as yet be completely computerized. These are markets that the educational system has been able to systematically provide for. However many positions are being grabbed by those more highly skilled labourers who have been displaced by technical advancements in machinery, thereby weakening the role of education in the labour market. Among those jobs that are difficult to be stereotyped include the supervision of other workers, childcare workers, cleaning and maintenance activities in the workplace, cooking, etc. Moreover, there are vast areas of work in those sectors excluded from the dominant economy, such as construction in households, street vendors, and others that can be considered difficult to displace. This all amounts to areas that can still be seen as a refuge for so-called manual labour, toward which a growing part of human labour is being shifted. This notwithstanding, even this area lacks sufficient reasons to be optimistic. As Frey and Osborne explain:

> Robots will likely continue to take on an increasing set of manual tasks in manufacturing, packing, construction, maintenance, and agriculture.

(...) As robot costs decline and technological capabilities expand, robots can thus be expected to gradually substitute for labour in a wide range of low-wage service occupations, where most US job growth has occurred over the past decades. This means that many low-wage manual jobs that have been previously protected from computerisation could diminish over time.

FREY and OSBORNE 2013: 22

As labour competition intensifies, wages fall. In the higher echelons of the labour force, the inverse can be visualized. Given that, in general, science as a productive force advances more rapidly than science as service (i.e., educational service), many of those who matriculate in educational institutions have to complete their practical training in the private sector. In this sense, the machine attracts human labour. An elite group is created in this way that grows very little socially while its economic importance skyrockets, alongside an uncontrollable growth of masses of workers whose economic condition inconsolably deteriorates.

But there is still more with respect to this problem. Brynholfsson and McAfee have found that computerization is being introduced in the field of non-routine tasks, or at least infiltrating the frontiers that had initially contained its penetration into human activities. Technology has begun to penetrate into areas that were thought to be inevitably destined to be completed by human labour. As they state:

Computers and other digital advances are doing for mental power—the ability to use our brains to understand and shape our environments—what the steam engine and its descendants did for muscle power. They're allowing us to blow past previous limitations and taking us into new territory.

BRYNHOLFSSON and MCAFEE 2014: 8

The observation of these authors is crucial and effectively defines what they call an "inflection point" in the development of society. While they amply document their argument, they don't see anything dramatically negative in this. Just like Keynes, they consider the undesired effects to be temporary situations that can be corrected if adequate measures are taken. They offer a series of short and long term policies in order to move the world forward towards prosperity, progress, and employment in a framework of shining technologies that these authors visualize for the future. Machines at present are for supporting the mind, as they lack intuition, creativity and the ability to respond in environments that are not predefined. The "inflection point" represents no threats for labour since machines only work to support living labour.

The ideas of Brynholfsson and McAfee as well as others have been deepened by Frey and Osborne who, while working on the technological development of the present, took up the task of evaluating how susceptible to computerization are routine and non-routine tasks, both manual as well as more cerebral ones. They established different levels of risk (low, medium and high) and concluded that 47 % of the total jobs in the United States are at high risk of becoming computerized within one to two decades (Frey and Osborne 2013: 38). This leads one to suppose that in a context of unstoppable technological development, these effects will be disseminated in a world that at the same moment suffers the need to create millions of jobs. Nevertheless, these authors are also not preoccupied about the developments placed into evidence.

How to Define the Present Situation?

On a daily basis, we read in the press as well as in the specialised literature about the truly prodigious scientific and technological discoveries that never cease to fascinate the world over. Just to mention a few:

- The press reported that three infants on the verge of dying from *tracheobronchial malacia*, a pathology that results in the collapse of the trachea and for which there is no cure, were able to recover thanks to implants created by a 3-D printer. The airway splints were finely crafted to adjust to the biological growth of the babies. This falls in the general line of prosthetics, hearing aids, etc., that is already well established. The 3-D printer appears to have no end to its capacity for generating prototypes in different activities.
- Computers have become enabled to analyse biopsies without risks of the possible errors that are frequently committed by medical analysts.
- Medical diagnostics are becoming computerized and it is already possible to individualize the symptoms, genetic trail and medical history, and even prepare treatment plans. Patients can also be monitored directly by computer sensors.
- Vehicles now exist that can operate without drivers, something that until recently would have been considered impossible given the number of human qualities that driving requires. It is not certain how some pending problems such as adaptability to changing climates will be resolved, but the devices have proven able to detect obstacles, evaluate positioning and observe the norms of traffic.

- Computers are able to surpass human talent in complex games, such as chess and others.
- Certain basic tasks for the design of statistical models are programmable.

The awareness that we are in the midst of a historically significant change is being extended in all spheres. It has not been long (25 February 2014) since the executive director of the International Monetary Fund (IMF) stated:

> 'Science fiction' is rapidly becoming 'science fact.' Indeed, we may just be getting started when it comes to the power and reach of 'machine intelligence.'
>
> LAGARDE 2014: 3

With respect to work and machines, three big moments can be recognised:

- The first moment was based in the Industrial Revolution and lasted throughout the larger part of the Nineteenth Century. The development of the machine rested in steam power. The introduction of cloth weavers, spinners, and mills had increasingly devastating effects upon employment. This resulted in fierce resistance as immortalized in the Luddite movement that took place between 1811–1816 in England. Ned Ludd had destroyed two textile frames in 1779, an act of defiance that inspired an anti-mechanization movement even if the true motives of this resistance have been debated. Machinery was then displacing the highly skilled labour of the artisan. This worker was only formally subsumed in capital since control over the rhythm of work and the mass and quantity of the product depended almost entirely on the arts of the labourer. The mental and physical activity remained united under the dominion of one person. Little by little, these two faculties of human labour became separated and the direct producer became increasingly dedicated to physical tasks that were confined to a position where submission to capital through real subsumption to the machine became evident.
- The second moment spanned the entire Twentieth Century. Mechanization grew with the expansion of electricity. Scientific work became increasingly organized and expanded through all areas of the economy. This did not, however, immediately displace employment. On the contrary, the initial introduction of this technology also corresponded directly to the expansion of employment in assembly line production such as in the Ford Motor Company. In this case, the principal objective was to reduce the amount of labour time in the production of a good, even as this signified contracting more

and more labourers. Industrial development, wars, the unveiling of mega-projects of infrastructural construction and the development of transportation systems took place and complemented the two forms of work, in spite of the advances registered in the computerization of routine work. Likewise being developed was a stratum of medium-skilled workers dedicated to the administration of companies and agencies.

– The beginning of the present period can be located in the 1990s. Scientific development and human labour are entering into a serious conflict. The entry of mechanization in non-routinized sectors signifies its penetration into what was formerly the dominion of brain work. Machinery is also displacing human labour in this area but at the same time, to the extent that it proves to be superior to human talent, it is likewise subordinating and subjecting people to its results. There are signs appearing of a really new form of subordination of human cognitive work to the tempo and results of smart devices.

It is important to establish that this discussion of technological development and its relationship to employment is not exactly explanatory. There is no technology that in and of itself produces unemployment and layoffs. As we well know, if technology has something to do with explaining these phenomena, it is due to its capitalist utilization. It is this capitalist utilization that has frustrated previous dreams of pleasure and free time as forecasted by Keynes and that will continue to frustrate the dreams of the followers of the present day model. What is certain is that science encapsulated in a capitalist context rarely serves any purpose not directly related to the generation of profits, a logic that immediately establishes the limits of its development. In general terms, the most relevant conflict at present is not with mechanization but with those that employ it for the purposes of exploitation and profit.

What more can technological development represent in this context if not the expanding surpluses of population and deepening social inequalities? Worse yet, it's not only the economic relations in the heart of capitalist society that are being reformulated as a result of technological progress. Greater unemployment and more social inequality will have an impact, which in practical terms is already occurring, on the political relations of society; concretely, on its democratic organization and on the relations between nations.

Social protest will tend to grow and along with it, repression and re-accommodation of state institutions in opposition to democratic guaranties. In the US, the documentation of these developments has gained some momentum in the context of the Occupy movement. Research on these issues has enabled us to observe the way in which the state is adjusting to these new conditions.

Jennifer Turner writes about how the policing institutions are reacting in the face of protest:

> The police possess strict guidelines for responding to protests: regulate permits in a pre-designed way to redirect, dissuade or prohibit protests; deploy large numbers of agents towards the areas of mobilisation; surround activists; use barricades to divert them or impede them from passing; create zones where protest is prohibited; detain people for minor infractions that are not usually penalized; and use force, including with "non-lethal" weapons, in order to control the situation.
>
> TURNER 2013: 9

That is the context in which the state creates new barriers and arbitrary powers that reinforce its position with respect to the larger society. The tragic events of 9/11, the nature of which will continue to be discussed for years, served as the excuse for a large number of laws and repressive measures that strongly colour the use of power in the United States. US attorney Jonathan Turnley (2012) offers a very good synthesis of these measures:

- The government can order the assassination of citizens considered to be terrorists or accomplices of terrorists, i.e., extrajudicial executions.
- People can be indefinitely detained as suspects.
- The state authorizes itself to decide if persons will be tried by civil or military tribunals.
- Suspected persons can be investigated without a judicial order.
- The government can use 'classified information' as evidence against suspects, including 'legal opinions,' i.e., judgements not necessarily validated by the law. This allows the government to invoke secret legal arguments to support secret processes using secret evidence.
- Obama, following the same policy as Bush, has insisted on no trials of CIA agents responsible for torturing suspects and has not allowed investigations of these practices to go forward.

None of this coincides with the image of a democratic state, much less of a country that defines itself as a nation of liberties. The image of a democratic state falls to pieces in the face of open assassinations of people of colour, including many Latinos among the most socially excluded sectors of that society.

The international situation, filled with militarization, armed conflict and threats of a wider global conflagration deserves a separate discussion, but certainly elevates the uncertainties of the moment. For a long time, it was thought that capitalist development offered the best possible promise of a society for employment and maximum freedoms. The most recent decades have placed this notion into open contradiction. As we have seen in the present work as well as in others, society is immersed in a wide-ranging conflictual dynamic that, far from realising the dreamed of ideals cited by the neoclassical thinkers, is instead being clouded over with the spectre of barbarity.

CHAPTER 3

Technology and Subsumption by Capital

Jesús Becerra Villegas

In the final writing of *Capital*, Karl Marx decided to postpone the discussion of the relationship between labour and capital in order to start his work on the study of the commodity form. He adopted an expository strategy and heuristic approach that was based on the study of those elements most immediately apprehensible to human consciousness, that is, the prodigious variety of commodities already being offered at that time by the capitalist market. Only after considering this phenomenon would he enter into the types of social relationships that underlie it and allow for the reproduction of the system.

A century and a half later, it is possible to find elements that on balance affirm the variety and scope of the various relationships established by the market. While this never invalidates the importance of the labour-capital antinomy, it does help explain much of what it is in everyday life and what forms it adopts. This approach presupposes something of a shift in the facts (or beginning from the facts) about what in the times of Marx already existed but which could be pointed to as only a phenomenological manifestation of capitalism. The sphere of circulation where market and commodities are located complemented in an indispensable manner a process dominated by production. Now the space of the former is much greater, exhibiting its own logic of reproduction. In other words, an arrangement of the world exists now that points to a historically positioned social mode that is sufficiently different so as to require a rethinking of the very nature of capitalism.

This work addresses the cultural dimension of the technically mediated relationships that sustains capital in its social environment, with the aim of subjecting the latter to the former's logic of reproduction. An exercise in descriptive exploration will be made of the role that proprietary technology plays in the so-called Third Industrial Revolution, or IR3. The aim is to understand the constitution of regimes that on the one hand produce the effect of "closure" or the completion of domination, while at the same time possibly undermining the sustainability of capitalism such as we have known it. To this end, we affirm the existence of a *logic* that serves as a naturalized mode of perception, valorisation, and actuation (Bourdieu 1990), i.e., an ordering of reason and proportions that defines the capitalist mode. This logic serves to build a passage of circulation in both directions between the social and the individual, between production and consumption, and between norms and practice. The naturalization of these movements that are today provided by technology presumably provides a human face for capital and its processes. Our analytical task must be understood to be located in the dimension of culture, the area for the production of a sense of belonging and community that has served the system in the past and which establishes for it a destiny.

The Social Formation in Digital Times

Marx proposed the Capitalist Mode of Production (CMP) as a logical historical category to describe the society of his era and perhaps of ours. Because the term is placed as an adjective, what is commonly called *capitalism* is just one more social case among others that are also historically situated. This analytical view calls attention to two elements: the modality as a form becoming the dominant logic and the productive factor as the embodiment necessary for the operation of this logic. Thus, the CMP can be understood as a concrete and complex formal entity, something that for the purposes addressed by the present work contains on the one hand relations in the wider social space such that it is defined by them. On the other hand, CMP is formed by materialities that, because of its nature or its effects, are capable of transformation or reproduction.

Maintaining the CMP for a period long enough to form a common epoch (that which was described by Marx and ours) where the productive relations of capital-labour are central, has an epistemological cost that rises along with the acceleration of processes, the provisionality of results, the cultural closeness of all which is consumable, and the proliferation and lability of supply. While the Marxist category of Social Formation (SF) allows us to sort out varieties of capitalism not only across geographic space but also over time, the complexity and prevailing uncertainty refute the categories of CMP/FS categories of thought in the same act as they are *explaining* those phenomena. We understand by *complexity* the densification that takes place at various social levels. By *uncertainty*, we refer to the effect of abandoning the protection of individuals, institutions, business corporations and government agencies. Social densification expresses an insufficient regulation for some as well as an excessive regulation for others, while uncertainty is experienced as the need for developing strategies of consumption in order to gain greater security.

Complexity as Densification

To think about present-day capitalism, it is possible to characterize the social formation that we are living as consisting of, in fact, a densification that is derived from four principal factors.

First, it involves an *acceleration* of the processes taking place in societies. The time taken for the development of new products and the innovation of new processes is getting ever shorter. More processes occur in increasing numbers of industries at the same time. Since many of these innovations result from monitoring existing consumption patterns, it can be said that they come forward to the market in a manner pre-appropriated by their users. This reveals that the regime has put into play a process of meta-appropriation that the consumer welcomes due to its confusion with their exercise of sovereignty (Becerra 2009: 46). The speed is also present in the capacity for the consumption of information *in real time*, but which does not permit the means for fully interpreting the complexity of data and the best ways of managing them (e.g., to form an opinion, decide on an action), all with some effect on the hierarchical position of the local setting.

Secondly, we have the provisionality or tentative character of the resulting acceleration of social processes that contributes to the above-mentioned densification by the rise in the rates of replenishment and consumption. The first is tied to the production of innovations and the replacement of products made with a pre-programmed expiration. In this context, an increase of consumption results from both a broadening out and deepening of markets, reaching more and more people to satisfy these manufactured needs in new and personalized areas. Packaging contributes through two movements that converge from different directions to the same point: on the one hand, through the dosing of products and their accessories, especially those with a rapid technological expiration; and on the other hand, a modularization that presupposes the purchase of replacement units on an ever widening scale that are ever less amenable to repair, above all in goods that would otherwise exhibit a long duration of effective use. These conditions pull at both the production of goods as well as at the formation of ideas and practices. In historical terms, they correspond to an epochal change marked by the emergence of patterns of finitude in social contracts that embody the terms that form families, jobs, affiliations or identities, such that brand loyalty emerges as the primary good that the consumer can sell.

A third source of social densification is the *cultural closeness of the object of consumption*. Likewise understandable as the effects of the market, new identities have been formed to generate a sense of community around consumption. When needed, this retakes elements originating from existing cultures so as to offer strategies of sociality in practices inscribed as the possibilities of goods. It is the generation of mechanisms of appropriation that at the same time give meaning to individual practices and also act as a means of satisfying a primary need, which is a sense of belonging and a subsequent closure of cultural distance. Nestor Garcia Canclini (1995) has associated the strategies of consumption with rationalities that can serve well as markers of social class and group affiliation, something that is otherwise inexplicable. So what we have described is a mechanism for densification that operates in more than one space for each individual. In other words, the consumer who belongs to a specific group, becomes complemented by various consumers of other groups he or she also belongs to in additional places, including virtual places. Thus, it happens that the market operates as an efficient producer of consumers or, seen from another angle, it can be said that the market is reproducing itself.

As a fourth point, *the proliferation and lability of capitalist supply* that circulates as a result of the state of affairs already described, constitutes in itself a confirmation of the immense power of the existing social mode and a mechanism close to its construction of power. If from an economic perspective, the CMP is presented as a logical historical social arrangement for the accumulation of wealth, from a more open view, the system reveals its orientation to individualistic consumption of all of its fruits of organization and social conduct. It is the sum of all those acts ordered for immediate fulfilment, or for which it is postponed while producing complexity and uncertainty, that is ever more ostensible in such a measure that they could be employed to chart our epoch. In the wake of the densification of material and symbolic supply, something ever more effective and likely to continue, the struggles to gain better conditions of access will be unleashed by diverse kinds of mechanisms on new and varied scales.

Uncertainty as a Need for Consumption

We have noted the existence of another greater dimension of the social configuration, whose weight should be implied in order to refer with the same meaning what today we call *capitalism*. To put this forward is not only possible when trying to envision a new mode of production (MP) or a stage of the current one, but above all when we wish to assess precisely whether capitalism is a mode of production. For this purpose, we must continue to devote attention to the *abandonment* that characterizes the social formation (SF) of our time, that which we have described earlier as produced by social densification. In one sense, perhaps it is enough to re-read that which we have said earlier, visualizing it rather as a *dis-facilitation* that the four sources of the social complexity produce in the performance of capital, the operation of governments and institutions, and the actions of groups and individuals. These latter actors become forced to shed those garments only useful during earlier stages in their history and to take up new struggles.

Given that social complexity results from the integration of subordinate and concrete forms that produce the emergence of other, more abstract ones, the cost of comprehending and controlling the complexity is greater than the sum of what is required to comprehend and control their generating parts. This means that those instances of lesser materiality, such as the financial and communications sectors, are better placed to gain immediate benefits from the states of emerging things than those that are more concrete, such as the primary and secondary sectors of the economy, government, institutions, groups and individuals. Proof of this is the role played by the complex in the outbreak of the financial crisis and the imbalances they created, with benefits accruing to the sectors of low materiality in charge of the activities already noted above. Precisely for them, uncertainty appears not only as a cost of living, but also as what investment strategies struggle for and direct towards regularities. If the complex is produced in part by technological development, the coping strategies for reducing uncertainty are likewise derived from the deployment and use of technology. With those strategies, the society in which we presently live renders the main lines with which the social formation is sketched in service and replication of the mode of production.

First, at the level of individuals and their groups, having already stated above the effect of appropriation in which they are the object of the act in which they play to appropriate themselves, certain technological developments can be identified along with their subsequent market strategies that favour *individuation*. PricewaterhouseCoopers (PWC), the network of legal service, analysis and audit firms, characterizes in the following terms the technological wave that already looms:

> The transformative potential of digital in the next decade is immense. In the consumer market, we think that this transformation will play out over three digital waves:
>
> – The First Digital Wave: mainly 'another channel' or 'e-commerce' through which to sell and communicate. This is already happening and even the most traditional businesses are adopting it.
> – The Second Digital Wave: digital will move beyond a channel to facilitate an economy of outcomes, where information shared through connected devices helps customers achieve the outcomes they care about.

– But we believe the Third Digital Wave is just around the corner and will be driven by consumers taking back their digital identity and extracting value from it. Whether in the form of 'buying brands' that aggregate demand across groups of consumers or life patterns providing platforms that will manage digital footprints and consumption.

Companies will find new value in this evolving marketplace by bringing together the four key aspects of digital: Social, Mobile, Analytics and Cloud. It will no longer be enough for companies to form a digital strategy to succeed in this world, they will need a business strategy fit for the digital age.

PricewaterhouseCoopers 2015

While this does not suggest an integral recovery of the individual, the plasticity of technologies for personal use will have to reposition the role of the consumer, marked with a history and aspirations for a niche in the market that capital wants to grow. Another aspect is the insertion of the subject and the groups to which he or she is attached in the spaces of real power, even in payrolls (Brynjolfsson and McAfee 2014). But amidst the contradictions of the system, there will be investments that need for subjects and groups to maintain and even increase their purchasing power. This tendency will be favoured by the steady fall in market prices for high-tech products to the individual consumer. For example, the cost of producing DNA sequences per genome was US$ 96 million in 2001 but fell to less than US$ 6,000 in 2013 (PricewaterhouseCoopers 2015). Finally, it should be noted that two forces are already present in this dimension. On the one hand, there is the "sharing economy" that separates the areas of property and consumption as well as individual interest from the interests of the community. On the other hand, the collaborationist culture, of which the various types of *wiki* forms, are an important kind of anonymous participation subordinated to the interests of groups they are part of.

Second, a shift of attention can be anticipated from the mere figure of institutions as entities, to the *institutional as culture*. In effect, it is worth postulating as a principle that *the best of a good society and the worst of a bad society are always their institutions*. Indeed, given that the institutional is an abstract modality that supersedes the level of individuals and groups as social entities, it is the institutionalization of social practices alone that can constitute a viable and integrative path forward in view of social densification. While the attainment of institutionality does not depend upon development and the appropriation of specific technologies, the strategies for achieving it even where culture does not provide institutionality, is favoured by the social availability of networked horizontal communication technologies. Moreover, the demands of a market once again operate in favour of this tendency by means of pressuring national economies to differentiate themselves in order to attract investments, this according to its institutional maturity as well as its participation in global exchanges and the size of its domestic product. Thus, PWC develops its portfolio and prospects from five dimensions: economic growth and stability, progress and social cohesion, communication technologies, political, legal and regulatory institutionality, and environmental sustainability (Pricewater-houseCoopers 2015). In the context of densification, it can be understood that the reduction of certainty has become transformed into a complex commodity of low materiality, high production cost and high valuation.

Third, the standardization that governments, especially national ones, can produce by using technologies of *assessment and surveillance*. The first is made possible by the juridical configuration proper to sustain a degree of control over the processes of production and for the transmission of knowledge in education systems, especially the universities. This assessment, which alleviates the burden of public administration for designing programs, is practiced by more and more governments that have found it to be an economical way to maintain control while sustaining the margin of play in autonomies. In the exercise of monitoring, the case of the US is a paradigmatic one as shown by John Bellamy Foster and Robert W. McChesney (2014). The authors analyse practices of monitoring, espionage and collusion of the US government against its citizens and corporations, both within and outside its borders, as well as against other governments and powers. They point to the role played by the media to sustain the war economy and mass consumption on a large scale by nurturing a public feeling of threat and manifest destiny, of containment of communism and for the defense of the free world. Then, with the repeal of the Glass-Steagall Act, the financial system was freed up, in the first instance, to produce monetary equivalents without material support, while at the same time its operations became rendered more complex and, by way of the newest technologies, financial products grew that produced periodic global crises, to the degree of constituting a "banking bankruptcracy" (García Domínguez 2015). This was bad business for the governments involved but magnificent for some of those who alternate between the roles of banker and public servant. Data mining and the invasion of personal, corporate and national communications are other practices that have already been reported and documented, which have cost the US government lawsuits and denunciations, but which also exhibit weak political institutions that must subscribe to and purchase access from the large companies of information and communications. Moises Naim also made this clear when he referred to the global existence of:

... a conflict characterized by the blurring of boundaries between war and politics, between military and civilian affairs. It is a conflict in which a violent non-state actor struggles against a state and in cases where the confrontation is military, not only in the strict sense of armed hostilities, but also because it develops between the media and public opinion, and because each side strives to undermine the foundations and legitimacy of the other so as to defeat the other on the battlefield. Terrorism, cyber warfare and propaganda are the common and habitual instruments of fourth generation warfare.

NAIM 2014: 179

Fourth, the strategy of productive capitalism to address the uncertainty arising from the increasing complexity constitutes the central focus of interest in this work and is discussed in the following section.

The Third Subsumption

Marx proposed as the ultimate objective of capitalism and the objective of every capitalist, the accumulation of capital. But what is the objective of capitalism when we integrate workers into it, or from another vantage point, consumers? While capitalism as a system becomes more complex, its living force appears as a drive to persevere and go for more, and it does so through its social agents. One can ask if there is some kind of principle inscribed in human nature that produces the capitalist idea like freedom > organization > satisfaction, the second of these is which the CMP portends to carry out as a means to reconcile the two values, albeit in an imperfect historical manner.

Although the principles of the CMP are not inscribed in human nature, it is historically the case that their allegedly natural character help form the reasons that justify it and make it resistant, adaptable and durable. Once capital is put in place as a means, there remains the problem of its realization. This can only be through human action: design, work, organization, and consumption. The remaining social practices and institutions are merely forms of representation, preparation or derivation of the four main dispositions for its reproduction. An important discussion left outside of the argumentation developed by Marx when he analysed the production process must now be undertaken with the aim of proposing not only the expanded reproduction of capital, but also its subsumption that is made of human actions, especially design, work, organization and consumption. It is in this sense that the exposition of subsumption offered in this paper should be understood.

In the unpublished Chapter VI of *Capital*, Marx develops a logical historical analysis of the subjection that capital imposes on labour in order to constitute itself as a mode of production. It becomes understood that this imposition involves a process of struggle whose net balance is the social formation that we today know and that in the dimension of interest of this exposition, we have characterized already. Subsuming labour under capital does much more that affect its purchase and dismantle its self-sufficiency of process and representation. It also involves the generation of conditions for a sustained and broadened organization of the transformative action of labour for the purposes of capital. This process which consists of a continuous uneven and abrupt development that remains in place from its earliest days up through the present, exhibits features from which it is possible to discern two forms or in two stages: formal subsumption and real subsumption. Before characterizing each, it should be noted that in both cases it signifies the domination of abstract (capital) over concrete (labour), that is, of the forms over the processes, until this social relationship becomes consolidated in that structure we call *capitalist* which governs the social formation. The structural domination over these social relations is supporting the notion of the abstract and the real, although they find themselves in distinct dimensions or modes of existence. Together they form a complex dialectical unity of super-ordination from the logical and from subordination towards the historical. As Marx states:

> What is generally characteristic of *formal subsumption* remains valid in this case too, i.e. the direct *subordination to capital of the labour process*, in whatever way the latter may be conducted technologically. But on this basis there arises a *mode of production*—the capitalist mode of production—*that is specific technologically and in other ways, and transforms the real nature of the labour process and its real conditions.* Only when this enters the picture does the *real subsumption of labour under capital* take place. (italics in original)
>
> MARX 2009: 72

Thus, specific or reproductive forms of work are designed (legally and technically) in concrete processes and organized (financially, productively and commercially) by capital for specific or reproductive consumption. Above these concretions, the asymmetrical capitalist relationship stands as a *mode* of doing, with all of its contradictions and as a marker of an entire era, upon which, we should remember, hangs the question of whether it remains in force today. Even in the face of the complexity and prevailing uncertainty, some authors (e.g., Bolaño 2000; 2013) consider the incessant emergence of new products, practices, associations, businesses and forms of representation as merely contemporary updates of the era and forms of real subsumption. However, and this is fundamental for the present work, the labours of design, organization and, in general, scientific and artistic creation which are the means for the real subsumption of reproductive work, *are today no more than formal subsumed work.*

As such, the reading being offered here proposes the existence of a new type of subsumption in capital, one that is not just of reproductive work, but is rather of the social totality, capable of forming self-representations aimed at consumption, as an investment in regularities and reduction of uncertainties. It should be considered that the second subsumption of labour under capital not only deepens but it rather qualitatively supersedes the formal subsumption of labour under capital. So the real subsumption not only subsumes reproductive labour, but also the prior subsumption! The passage from the first subsumption of labour to the second is not a turnabout but rather an extension of the former. The legal market domination of a moment is not transformed into another, technical one, but instead adds to it as an extension of the former and historically produces the CMP.

The dominated class sustains a first loss that deepens without recovery in the second. Nevertheless, in the analysis of these stages, what has been missing is to note that when installing the real subsumption of labour, this is what becomes dominated and not the representations of the working class, which itself remains potentially revolutionary. What is needed is a symbolic subsumption that other forms of capital are responsible to undertake with the aim of clamping shut the space of subjectivities that escape the structured processes of production and are installed in the processes of consumption. This is not simply capitalist produced goods, but is rather a grand narrative that sustains capitalism, the narrative of freedom and the satisfaction of certainties through consumption. It is not just the cultural enterprises that are responsible for producing the imaginary that accompanies the reproduction of the system, or the practices of the financial system. Nor are only the academics charged with putting into circulation the necessary logic. But it is rather the space and rhythms of life in capitalist scenarios that become converted into massive showcases. Cities, roads, institutions, laws and procedures, all form a convincing supply to be savoured in the acts of micro consumption. The same occurs to narratives (Barthes 2002), the feeling of each act of consumption is not found in the act itself but rather floats overhead.

The third subsumption, no longer of labour but of culture itself as a logic of reproduction of the system, finally displays the weight of the subjective and aesthetic component in the mode of production increasingly defined by consumption strategies. As characterized by Fredric Jameson (1991), postmodernism becomes understood as the logic of late capitalism. To this, we must add that early capitalism produced, as well as industrial revolutions, a subsumption of culture and communication. As such, it can be proposed that the existence of a mode of culture and communication was in force and capable of naming a capitalism that presents contradictions that seem easier to resolve in actions rather than by conceptualizing, except when we stand from a new and different proposed subsumption.

Let us begin from a historical reading in order to place the construction of capitalism as it unfolds today in terms that we know. Robert J. Gordon (2015) located industrial revolutions (IR) in the following terms: the IR1 came about by the introduction of steam engines and the railroads, from 1750 to 1830. The IR2 was based on electricity, internal combustion engines, domestic running water, home health services, communications, the entertainment industry, and chemical and oil industries, with its peak period from 1870 to 1900. The IR3 consisted of the appearance of computers, the web and mobile phones; beginning in 1960 and continuing to this day. Gordon contends that the IR3 can no longer have the same impact on production processes or standards of living, as did the IR2. He argues that the new industrial revolution has introduced few substantial changes and that those changes since 1970 have been more in the way of "second round improvements" (Gordon 2015: 13). In our analysis, formal subsumption or the first subsumption of labour prepared the conditions for the IR1 while the second subsumption paved the way for IR2. As an additional effect, changes in the lifestyles of the population resulted in a formal symbolic subsumption of culture to capital, noting that it is still a formal mode in that it involved the generation of conditions of supply for consumption while still being governed by mass designs. The IR3 roughly coincides with the actual symbolic subsumption of culture to capital, in which the paradox occurs of a deep domination of consumption via the liberation of the means for individualized or group exercise of reference or collaboration.

This issue is of key importance because it explains how even with the slowdown in productivity growth of the IR3 as compared to the IR2, and moderation of the curve of increasing salaries related to work, the system has managed to maintain its profitability not at the cost of destroying capital, which is what existed prior, but through the deepened social availability of supply. Differentiation, aging and the de-massification of consumption translate into an effective and consistent extension of the market not only by adding more buyers, but also in the multiplication of each one of them through more acts of acquisition. This is due to the widening patterns of consumption, the decline of the amount of labour required to earn the purchase price, and the increasing rate of replacement of the goods being purchased. In short, it is a subsumption that provides the necessary culture of consumption for the system, allowing it to sustain this above the decreasing value of wages and the transactions of individuals in each branch and capitalist firm. Our societies have entered into a stage of collaborative consumption rather than into an epoch of production in the socialist sense. Automation and extreme specialization of labour processes as mechanisms for the accumulation of one class rather than for the consumption of all, makes unlikely collaborative work as a mere result of the expansion of capital.

Therefore, the increase in the volumes of consumption which are well above the rate of population growth in spite of the decline of real wage earnings and relative to the salaries of executive positions and the returns of shareholders, points to a capitalism as a structural and cultural mode of consumption that "pulls" capitalism as a mode of production. While technologies displace human activity from production to consumption, they confirm it as the *raison d'etre* of the direct producer. It should therefore be put forward as a kind of economic Pareto optimality, consisting of an "overweight" consumption in excess of the weight of production. In this regard, we should not forget that consumption is also of capital, especially for the financial sector in this case. The arrival of this optimality strengthens capital, although not necessarily the productive type. Capital that lives to move the boundaries between production and consumption towards the optimization of the system is that which carries the highest return on the state of things, especially given that its market, its clientele, will be precisely that which corresponds to the new regime.

Our reading suggests that the third subsumption is an epoch endowed with the logic of an increasingly growing complexity to which we must add one last point. What brought on the third industrial revolution (IR3) is a larger change of paradigms, one that affected not only the perception of the environment, but self-perception and self-esteem. It created referential communities for consumption, valuation, opinion and political practices. These new material conditions bring their own logic of reproduction that supposes a new narrative from previously unknown codes. Just as every media outlet, following its invention, had to mature in order to create its own language, aesthetics and unique type of stories, this new regime must do so, facilitated by consumer technologies that have been generated in its logical state of *real* subsumption of culture, the third subsumption, in order to circulate accordingly. In this sense, technology itself has become not only a set of technical methods, but also a culture of the epoch that younger generations especially overflow.

Conclusion

Although this exposition ends here with the emphasis on the cultural component and consumption, its inattentiveness to labour-capital relations is not an attempt to dismiss it, but rather to explore how the *market dimension* has gained in explanatory power in most recent times. In any case, for the sphere of production, it was seen as important to distinguish reproductive labour from creative labour, such that the latter is unconquerable in an absolute manner by capital, because to do so would always suppose a higher level than that which organizes its real subsumption. But the final emergent level can only in the formal sense be organized and purchased in a capitalist manner.

So we have proposed a reading that without claiming to provide a definitive answer on what today is capitalism, has established the question about the essence of it rather than taking it for granted. With this intention, we consider the market effect of the relation:

Production / Supply > <Demand / Consumption

Market, as language does, conceals itself by showing off. Thus, the face of consumption that the capitalist sees is demand to which it is oriented, while the face that the consumer knows is that of supply. In the midst of them stands the market as a mechanism of immediate and direct everyday contact of everyday activity:

Production / Supply > Market <Demand / Consumption

Our approach is valid for all expressions of the mercantile mode, as markets are concretized in merchandise. It is that which is produced and brought to market, that which is the object of demand and which is consumed, all of which satisfies the needs of each pole. For the purpose of analysis, it seems reasonable to posit the market as the manifestation that assumes capitalism or, in other words, as the representative *form* of capitalism. But in spite of the fact that the market is known to be blind, short-sighted and selfish, capitalism as a higher abstraction consists, above all, of a logic that does not need to see, project or be fair in order to be sustained and extended over time. It is only required to turn into a mode to see, value and actuate in order to exist in particular acts and institutional settings. In this sense, capitalism as a device constitutes in itself its own technology placed into action in order to subsume the world. This is what is now happening because the labour-capital relation is mediated by a subsumed imaginary, such that in the end, all of the technologies that capital produces are technologies of subsumption.

CHAPTER 4

The State and Freedom of Public Network Space

Sergio Octavio Contreras

The Internet was born out of a free culture. This is affirmed by at least two historical studies on the origin of this new technology (See Abbate 1999; Veà-Baro 2013). From this perspective, the Internet was the result of collaborative work and shared knowledge. The way in which it was configured, up to the present, was to allow a measure of freedom for users to interact with other users, to share information, design and content.

Currently there is much debate about the effects being generated by this electronic interaction and the way it configures new spaces of socialization. Pierre Levy (1994) argues that the Internet allows the union of minds, as a kind of shared brain where the sum of the individual intelligences can generate a collective intelligence. A different view is held by Derrick de Kerckhove (1997) who argues that more than a collective intelligence, what prevails in the information age is *connective intelligence*. People enrich their intelligence by interacting with others and exploring different points of view. Collective or connective theory shares one thing in common: people are joined in networks that hover above time and materiality. Individuals who are linked to other individuals to share information and to work collaboratively to build knowledge do so because it is in a place where everyone can participate and where hierarchies are diluted. The network is therefore a horizontal field of communication, non-cantered, where what is exhibited becomes the knowledge of many. This new sphere has three characteristics of public space: it is common, it is visible and it is open (Rabotnikof 2006).

Some studies suggest that new technologies cause new problems of contemporary modernity (Girola 2005) such as the strengthening of individuality to the detriment of community (Pardo and Noblia 2000). Other scholars argue exactly the opposite, namely, that electronic networks promote community (Wellman and Rainie 2012) and make possible a new form of sociality based in advanced communications (Castells 2006). What has been demonstrated empirically is that within prevailing organizational forms, the use of networks has changed the traditional structures through which people used to socialize with others, such as the spaces of distribution and manner of discussing issues of public interest. This change has resulted in an arsenal of new civil actions ranging from virtual protest letters to flash protest marches and agile street blockades that are network organized. Some of the detonators of protest originated in revelations of the violation of laws or moral codes of conduct by people in positions of authority, or other prominent personalities.

One such case occurred in May 2013 when Edward Joseph Snowden, former technician for the CIA and collaborator with the National Security Agency (NSA), leaked to *The Guardian* and *The Washington Post* news of a clandestine surveillance program over the public Internet (Greenwald 2014). The disclosure revealed that US intelligence authorities had spied on the lives of millions of Internet users worldwide, including presidents and political leaders. Another incident occurred in 2014, when emails sent by the governor of New Jersey, Chris Christie, were widely distributed. Those emails detailed a plan to intentionally create roadway traffic congestion on the George Washington Bridge that connects Manhattan to the New Jersey city of Fort Lee. A closure of two lanes caused heavy traffic and among other things resulted in the death of a 91 year old woman on account of her being detained from obtaining urgent medical care. The action was taken due to a local political dispute involving the mayor of a neighbouring New Jersey town, Democrat Mark Sokolich, who had earlier refused to support the Republican governor Christie in his re-election bid.

In Mexico, the violation of established codes was visible on the Internet also. In March 2015, neighbours of David Korenfeld who was then director of the National Water Commission (CONAGUA) took photographs and videos which showed that he and his family were being transported in an official helicopter from Mexico City as they embarked on a vacation to Colorado. The images generated a Twitter campaign entitled #RenunciaKorenfeld (Resign Korenfeld). After a week of pressure, the official was forced out of office. These cases produced their respective results on account of two factors, namely, that the Internet is a public space with a broad capacity to publish information, both public and private (Bauman 2006), and that the use of individual channels, such as blogs, social network pages, messaging services, applications for smart phones, etc., all form part of an opulent informational panorama (Carr 2011) that on many occasions can even provide the foundation for networks of untruths.[1]

---

1   See Sergio Contreras (2013).

Control of Public Networks

The free flow of information across electronic networks can be a problem for those who prefer that certain information should neither be known nor seen through these new media. In recent years, governments and power groups have tried to erect dikes against the flood of common and visible information that is available across the Internet. All such initiatives to restrict network activities invariably resulted in direct clashes with the inhabitants of these virtual communities. During the last decade, a review of the actions taken to limit electronic networks shows that three general mechanisms have been used by the State with the support of technology companies to suppress network use. They include both legal and covert illegal forms of coercion:

– Measures to control meanings (ideological, religious, cultural, etc.)
– Measures to regulate the market (protection of private property, intellectual property, patents, etc.)
– Measure for policing (to protect public safety, to combat terrorism, attack crime, etc.)

These measures invariably are applied as *Tools of Control* (TC). Such tools (Table 4.1), which can be institutionalized or not, usually operate in two ways: as regulatory TC (for example, laws or regulations) so that society fulfils them; and as communication infrastructure technical TC (for example, interception of messages, the use of automated programs to eliminate or block content to electronic web content). In the face of these TC measures, cyber navigators have had to make use of alternative technologies designed by the users themselves in order to evade the controls being imposed upon them. According to a 2014 report on worldwide Internet freedom that was compiled by Reporters Without Borders (RSF), both governments as well as security and surveillance companies have become the main enemies of social network freedoms.[2] The report identifies thirty-two companies and public agencies as new enemies of social networking freedom. Included in the list are the intelligence agencies of the United Kingdom and the United States as well as the public policies of some Asian and Middle Eastern governments.

---

2 Reporters Without Borders. *Enemies of the Internet 2014*, RSF. [Accessed 20 March 2015]. Available at: http://12mars.rsf.org/wp-content/uploads/EN_RAPPORT_INTERNET_BD.pdf.

TABLE 4.1   *Tools of control*

Type	Field of application	TC legal/illegal
Meaning Control Measures	Freedom of Thought	Agreements/Treaties Laws/Regulations Technology/Cyber Codes
Market Control Measures	Free Culture/Open Access to Knowledge	
Policing Control Measures	Collective / Private Life	

Based on data from the OpenNet Initiative laboratory at the University of Toronto as well as the Centre for Internet and Society at Harvard University, we can reaffirm that the TCs are part of a regulatory system set up by state agencies whose purpose is to restrict certain freedoms of cyber users, ranging from the freedom to choose to consult certain kinds of information, to the possibility of establishing links with people living outside of certain physical territorial boundaries. In the majority of cases, regulations seek to impose limits on areas related to the political sphere, i.e., ideological protest and social mobilization. Or they seek to impose limits on the freedom of information or the right to know, through open censorship of data flowing across networks or which are archived in some virtual space. The control that governments seek to exercise through the deployment of such mechanisms is most often presented as a sanction, i.e., the imposition of economic or legal punishment against offenders.

Meaning Control

In 2010, the government of Pakistan blocked the general population from using services provided by YouTube, Wikipedia and Flickr on the basis that these providers promoted blasphemy against the Muslim religion. Such measures had prior precedent. A group of cartoonists opened a Facebook page with more than 45,000 followers under the name "Draw Muhammad" where cyber users were invited to draw images of the prophet and place them on the site. Users managed to publish more than twenty-eight thousand drawings before the Ministry of Information Technology requested that the Telecommunication Authority of Pakistan block access to the page by following an order issued by the High Court of Lahore. The conflict eventually resulted in street mobilizations where thousands of believers protested against Facebook and death threats were made against the cartoonists (Lykkegaard and Westergaard 2012).

In the Ukraine, the Facebook account of Mykola Sukhomlyn (see www .facebook.com/sukhomlin) was eliminated by the administrators of the social network in June 2011. This resulted from the publication of a video that Sukhomlyn posted on 17 May 2011 where the governor of the Donetsk region, Anatoly Bliznyuk, appeared driving a luxurious Mercedes Class 2 valued at more than 60,000 Euros. The video was distributed by the digital newspaper *Ukrainska Pravda* and shared by thousands of Internet users, prompting an outpouring of public criticism against the governor.[3] According to the activist website *Maidan.org*, Yeketarina Skorobatova who works with a social media site in Russia reported that the elimination of the Facebook account that carried out the exposé was carried out allegedly because it had engaged in copyright violations by stealing jokes from another page *Durdom.in.ua*.

Alexander Aan, an atheist Indonesian, opened a Facebook page in January 2012 that questioned the existence of God. In a message circulated by social media networks, he wondered "If God exists, why do bad things happen?" The comment came to the attention of the highest hierarchical body in Indonesia, the Council of Ulemas. Aan was denounced to the police for blasphemy and inciting hatred. Based on the Law on Information and Electronic Transactions, he was sentenced to two and a half years in prison and fined more than US$10,000.[4] In 2013 the Chinese government, through a prosecutor and a high court ruling, filed a legal interpretation by which a person who spreads false rumours on the Internet that become repeated more than 500 times can be charged with defamation and receive a prison sentence. The restriction was put into effect one year after the arrest of Dong Rubin, a popular blogger who utilized social networks to report cases of police brutality in China. The cyber activist was sentenced to more than six years in prison and fined US$56,000.[5]

In 2014 the Syrian government blocked Skype and Facebook services. According to a study by the research centre Information Communications Technology of Australia, the Syrian regime used the most sophisticated control system machines to filter traffic information circulating on the Internet, using proxy servers manufactured by the US company Blue Coat. A proxy server is a network that acts as an intermediary between requests that a client or a

---

3 Internet Freedom Organisation. "Open Letter" [Accessed 8 May 2014]. Available at: http://
   uainfo.org/heading/public/2917-otkrytoe-pismo-k-marku-cukerbergu.html.
4 See Cochrane (2014).
5 See CHRD (2014).

server makes to another client or server. If an Internet user (A) from a particular machine sends a request to access a resource (D), then to reach such content, the request must go through the points B and C as part of the operations of an electronic communications network. When an Internet user in Syria requests information through a page like Yahoo!, servers that are under the control of the government can decide whether to grant access or not.

In other countries like Turkmenistan, Vietnam and Bahrain, part of the network infrastructure itself is controlled by the regimes. On July 2014, hundreds of youths began protests in Hong Kong against the government after it became known that public officials running in the elections of 2017 would have to have been appointed by the Communist Party. The protesters, mostly based in the universities, created the #OccupyHongKong movement through which they erected an Internet platform called *PopVote* in order to consult people as to whether they preferred free elections or designates from the state. The movement was demonized by local media such as *The Global Times* and the Xinhua News Agency and became the target of cyber-attacks. The movement responded by organizing street demonstrations enhanced by the utilization of digital networks. In September of the same year, the government moved to cut access to Facebook, Twitter and Instagram, while deleting running comments on *Weibo* (www.weibo.com), the most popular digital social network in China.

The young protesters of this movement proved to be resilient by using their mobile phones and the application *FireChat* to evade the electronic blockades being launched against them. According to the research portal *Wiboscope*, a site that studies from between 50,000 and 60,000 messages published daily in China, the greatest censorship throughout the entire history of *Weibo* was experienced during September 23–28 of 2014 during which time the government managed to eliminate millions of messages from the platform through the deployment of an army of programmers and specialists.[6]

According to the study Freedom on the Net: 2013 conducted by Freedom House, those countries with the greatest restrictions on social freedoms in the world are Burma, Sudan, United Arab Emirates, Belarus, Pakistan, Saudi Arabia, Bahrain, Vietnam, Uzbekistan, Ethiopia, Syria, China, Cuba and Iran. Mexico was placed in the category of partial freedom, sharing the same level as Angola, Uganda, Indonesia, Tunisia, Morocco, Libya, Jordan, Bangladesh and India. The report includes details on twenty-four countries that have passed laws against freedom of networking and thirty-five nations and institutions

---

6   Wiboscope. Censorship Index. [Accessed 12 December 2014]. Available at: http://weiboscope .jmsc.hku.hk.

that have increased their active monitoring of cyberspace. The ten mechanisms identified in use during 2013 for curbing the free circulation of meanings were:

1. Content blocking: in South Korea, Russia and Jordan political and moral issues are constantly being censored.
2. Cyber-attacks on dissidents: Malaysian portals or social activist networks are blocked by distributed denial-of-service (DDoS) attacks.
3. Laws against Internet users: citizens were arrested in 28 countries for using electronic networks to express political, social or religious judgments.
4. Paid Internet users: Bahrain and Belarus used an "army" of online commentators to attack regime opponents.
5. Physical assaults and murders: actions ranging from threats and kidnappings to executions against bloggers and activists have been recorded in Syria, Egypt and Mexico.
6. Espionage: there is an escalation in the use of new technologies by authorities to monitor those not politically aligned with the government in power.
7. Elimination of content: it is increasingly common that various spheres of power request that companies remove content for violating certain interests.
8. Locking down of networks, applications and communications: it was notable that the use of Twitter was banned in 19 countries.
9. Penalties against intermediaries: laws have been enacted in 22 countries to punish Internet service providers.
10. Shutdowns: some governments, such as India, Syria and Venezuela, have shut down the network infrastructure during moments deemed to be sensitive.

Control of the Market

During the first decade of the opening of the Internet, innovations for sharing information were collectively created. Such was the case for electronic tools designed to distribute files such as music, videos or documents. This is how *Napster* arose in 1998, a service that stored millions of songs and allowed visitors to download them in an audio format known as MP3. This media format was designed by the group *Moving Picture Experts Group* (MPEG) and standardized in 1995 under an algorithm (pre-set instructions for performing a transfer) that reduces the volume of data contained in an audio track.

A year after its launch, several record companies sued the service for the storage and distribution of music without having previously purchased them. In 2001, *Napster* was shut down by a court order. Appearing in their place were new forms of free distribution such as *Audiogalaxy, Morpheus, Kazaa, LimeWire* and *Gnutella* platforms. These programs allow a document already held by a user to be shared with others. There was no longer any physical space being used to store files like in *Napster*, but instead, an active collaboration of millions of Internet users who were sharing parts of files. After the closure of Napster and other programs, the use of *BitTorrent* became popular through which one could download a movie or a complete collection of disks through the use of "peer-to-peer" data sharing technology. This protocol allows anyone trying to share a file to first convert it to torrent format and then distribute it to storage servers. Those interested in obtaining the file can then begin downloading and simultaneously distribute it to others who also want it. Some popular programs that are based on this system are *Vuze, KTorrent* and *BitComent*. Under this system, *The Pirate Bay* appeared in Sweden in 2004 and became one of the most popular torrent search engines of the decade.

In the face of these technologies of content distribution, the State has been concerned with trying to regulate these areas when free distribution is considered to be damaging to the market.[7] A study by the company Business Software Alliance (BSA) indicates 59 billion dollars are lost each year from the use of "pirated" software and the illegal download of Internet applications. According to the Global Piracy Report, those countries that most contribute to this phenomenon are Brazil, China, Russia and India. In the United States, about 20% of software is "stolen" while in China, it is closer to 50% and even 65% in Russia. Overlooking these practices, the excuse of some governments is to protect the neutrality of electronic networks, i.e., the transmission of data flowing from a source to a receiver (Alcántara 2010).

Regarding laws to combat such practices, Spain approved in 2009 the "Sinde Law" which includes stringent measures to protect intellectual property on the web, mainly films and commercial music recordings. It authorized the Ministry of Culture to close any website that according to their criteria is affecting property rights of third parties. The name of this law passed on 27 November 2009 by the government of Jose Luis Rodriguez Zapatero was in honour of its sponsor, Angeles Gonzalez Sinde, who was Minister of Culture. A similar law was enacted in Colombia by Interior Minister Germán Vargas Lleras and was

---

7   Business Software Alliance. "2010: Piracy Study." [Accessed 18 June 2013]. Available at: http://globalstudy.bsa.org/2010/.

approved on 4 April 2011 as the "Lleras Law." This legislation authorized the suspension of network service to electronic network users that had profited from the content of intellectual property they had not properly acquired. The protests against this law were so potent that the government had to back track on the law, leading the President of the Colombian Senate, Juan Manuel Corzo, to declare on 16 November 2011: "Today I promoted the collapse of a law that curtails freedom of communication."[8]

The governments of Canada, Japan, South Korea, Morocco, New Zealand and Singapore, signed with the United States in 2011 the Anti-Counterfeiting Trade Agreement (ACTA). This agreement had the support of major multinational financial institutions such as the Recording Industry Association of America (RIA) and the Motion Picture Association of America (MPAA), who represent music production companies and major film producers worldwide. ACTA is an international trade agreement to which governments concerned about protecting the rights of companies against counterfeiting and piracy, both physical and digital, can adhere to. On 5 July 2012, the 27 countries represented in the Parliament of the European Union voted 165 votes against and only 39 in favour, thus rejecting the entry into force of ACTA within the EU.

In the United States, an initiative entitled "Preventing Real Online Threats to Economic Creativity" or (PIPA) was introduced in May 2011 in order to punish the creators of websites that freely share video, music, movie or other production content in violation of copyrights. After continuous protests in cyberspace and criticism of the US government's stance, the Senate Majority agreed on 18 January 2012 to postpone a vote on PIPA. Another similar initiative was the Stop Online Piracy Act (SOPA), also known as Law H.R. 326v, filed in the United States House of Representatives on 26 October 2011. The SOPA was supported from its inception by more than 350 US companies. The main argument advanced by its proponents was that economic losses were being perpetrated by the large transnational communication, estimated at $250 billion annually, because of downloading free content across electronic networks. The initiative included sanctions such as the closure of web portals, the removal of links within search engines, and sentences of up to five years imprisonment for those found guilty of violations. In January 2012, more than 10 thousand websites were "turned off" as a means of protest and the White House backed off of its support of the initiative.

---

8   *El Tiempo*. 2011. "Archivada la Ley Lleras, que pretendía regular el uso de Internet". *El Tiempo*, 17 November.

Perhaps the best known case in exchange content censorship was the closing of *Megaupload*, a sharing service that had grown to 150 million subscribers. The Federal Bureau of Investigation (FBI) blocked the site on 19 January 2012 and arrested its executives, charging them with illegally amassing $175 million in profits at the expense of $500 million of copyright losses. The founder of *Megaupload*, Kim Schmitz who is known in virtual network circles as Kim Dotcom was jailed for several months in New Zealand along with three of his collaborators. Upon leaving prison in 2013, he announced a new service name *Mega* that encrypts files to prevent programs designed to track file downloads on the part of companies trying to detect copyright violations.

Police Control

A third way in which the State intervenes in the free use of electronic networking relates to law enforcement mechanisms aimed at guarding public security. Among the pretexts for court interventions are included the war on global terrorism, the struggle against drug cartels, the pursuit of dangerous criminals, the prevention of transgressions, and the policing of crimes that fall under special codes and regulations, such as child pornography, identity theft, illegal arms sales, cyber fraud, and others.

On 30 November 2011, the Cyber Intelligence Sharing and Protection Act (CISPA) was introduced in the United States Congress with the support of numerous American politicians and corporate developers of electronic technologies such as Intel, IBM, Oracle, Microsoft and US Telecom, among others. CISPA allowed police and espionage agencies to intervene in the digital communications of users when there was a suspicion of terrorist or criminal activities. The proposal, even though approved by the US House of Representatives, did not pass the US Senate and was openly opposed with a veto threat by President Barack Obama.

In early 2012, a group of hackers from the Chaos Computer Club (CCC) accused the German government of Angela Merkel of spreading a Trojan horse type virus that acts as malware and records the computer registers of network users. The CCC[9] showed how the Trojan infection was spread from government offices. The F-Securite company confirmed that the virus in question was a

---

9   Chaos Computer Club. Publikationen, "Analyse einer regierungs-malware", [Accessed 14 January 2015]. Available at: http://www.ccc.de/system/uploads/76/original/staatstrojaner
    -report23.pdf.

program that operated as a keylogger, i.e., it records the actual movements on a keyboard, makes screen shots and records audio. The German government denied the allegations but acknowledged that the program existed and has been used under court orders to stalk network users considered suspects of criminal activities.

In September 2013, the WikiLeaks website released a report known as "Spy Files 3" which consisted of documents that recorded the trips made to various countries by representatives of three tech surveillance companies, namely, DreamLab, Gamma and Hacking Team. The report also included a list of 190 companies that sell their services to various governments in order to monitor people who use the Internet.[10] In February 2014, the Citizen Lab project, led by researchers Bill Marczak, Claudia Guarnieri, Morgan Marquis-Boine and John Scott of the University of Toronto, presented a report in which it was revealed that 21 governments of different countries bought the company Hacking Team application Remote Control System (RCS) that was capable of deciphering passwords, intercepting electronic communications and planting Trojan viruses for the purpose of data collection. Among the nations that acquired this technology were Egypt, Hungary, Kazakhstan, Korea, Malaysia, Nigeria, Oman, Panama, Poland, Saudi Arabia, Thailand, Turkey, United Arab Emirates, Italy, Uzbekistan, Colombia and Mexico.[11] According to Citizen Lab, the RCS has been used in Ethiopia, Morocco, Uzbekistan, Saudi Arabia and Sudan to attack opposition journalists. In Mexico, it is carried out through five different phases, first in the country where the information is obtained, then it moves to Hong Kong, then on to Atlanta, Amsterdam and finally to London. Through these covert servers, it is possible to avoid leaving traces of the activity.

On 29 April 2014, Vlora Citaky, the Kosovo Minister of European Integration, posted on her Twitter account a message through which she announced the submission to Parliament of the Interception of Communications Act.[12] The legislation requires service providers to give the government full access to information about any electronic network user who is suspected of committing

---

10   WikiLeaks, Spy Files, "Spy Files 3", Spy Files, [Accessed 2 August 2014]. Available at: wikileaks.org/spyfiles3.

11   University of Toronto. Citizen Lab, "Mapping Hacking Team's 'Untraceable' Spyware", Citizen Lab, [Accessed 11 May 2014]. Available at: https://citizenlab.org/wp-content/uploads/2015/03/Mapping-Hacking-Team%E2%80%99s-_Untraceable_-Spyware.pdf.

12   Republic of Kosovo Assembly. "Draft law on interception of telecommunication", Republic of Kosovo Assembly, [Accessed 4 March 2015]. Available at: http://www.assembly-kosova.org/common/docs/ligjet/04-L-173.pdf.

any crime, or intends possibly to commit one in the future. In August 2014, the government of Australia put forth a proposal to modify the operations of the Australian Security Intelligence Organisation (ASIO). This reform would enable the police to add, delete, copy and modify any data saved on any computer that is connected to the Internet as part of an ongoing investigation.[13] It also provided for up to 10 years of imprisonment for people who disclose government intelligence operations across the Internet.

The Case of Mexico

In Mexico, all three forms of cyber control described earlier (symbolic, economic and coercive) have been used by the state to attempt to regulate the transmission systems of Internet messaging. In recent years, political authorities proposed to legalize the mechanisms by which they can firmly control some key sectors of online networks. This has mostly followed paths taken abroad, especially US policies that wield an economic vision and aim to criminalize a significant portion of cyberspace. Since 2010, the interests of the Mexican state to establish virtual network controls have revolved around two points: 1) cyber laws (Tellez 2004) that include issues regarding intellectual property, the use of personal data, informational data security, and cybercrimes; and 2) political laws that address social participation in electoral contests, freedom of expression and the right to know. Some examples of regulatory policy instruments aimed at the network that were registered in Mexico over recent years are outlined below.

In early September 2011, the federal deputy of the Institutional Revolutionary Party (PRI), Arturo Zamora, presented before Mexico's Congress an initiative to reform the Federal Penal Code in order to provide up to nine years imprisonment for those who utilize electronic networks in order to defame candidates, political parties, institutions and public officials during election periods. After a protest campaign mounted across Facebook and Twitter, Zamora withdrew his initiative. In the same year, the governor of Veracruz, Cesar Duarte, promoted and passed a law to criminalize rumours that were spread by social networks. In June 2013, the Mexican National Supreme Court of Justice declared that the "Duarte Law" was unconstitutional because

---

13   Inspector-General of Intelligence and Security. "Annual Report 2013–2014", Inspector-General Intelligence and Security, [Accessed 5 November 2014]. Available at: www.igis.gov
.au/annual_report/13-14/pdfs/IGIS_annual_report_13-14.pdf.

it was founded upon violation of the freedom of expression as well as violation of the privacy of individuals.

During the years 2011 and 2014, similar initiatives were taken by the local congresses of Tabasco, Nayarit, San Luis Potosi, Queretaro and Zacatecas. On 15 December 2011, a Senator of the National Action Party (PAN), Federico Döring, presented two proposals known as the "Döring Law"[14] that aimed to amend some of the articles contained in two of Mexico's national laws: the Federal Copyright Law and the Industrial Property Law. The law sought to hold Internet users responsible to compensate companies who owned content being freely shared across networks. Those who refused to comply would be called to appear before the authorities and have monetary fines levied against them. Thousands of network users organized a campaign of protest against this proposed law with the hashtag #Ositodoring. This campaign gained the support of the group *Anonymous* whose online activists initiated an operation on 27 January 2012 at 11 am called #OpDoring that adversely affected the operation of various electronic portals of the Mexican Senate and the Interior Ministry. Cyber activists likewise revealed the passwords of email and cell phone users of various political figures who belonged to the PAN, Doring's political party.[15] The international movement Avaaz.org collected more than 42 thousand digital signatures against the proposed law. Ultimately, all of this pressure and criticism mounted by the local and international social movement organizations led the Mexican Senate to "freeze" the proposed law in its tracks.

In late March 2012, legal reforms to the Federal Legal Code were approved in areas regarding "cybercrime," specifically Articles 205, 211, 282, 389 and 390. The reforms as presented by a deputy of the Environmental Green Party (PVEM), Rodrigo Perez Alonso who was serving as secretary and presided over the Digital Access Commission, established penalties of three years imprisonment for the wilful act to alter, destroy or cause the loss of information contained in computer equipment of the Mexican state. On 11 July 2012, the administration of President Felipe Calderon signed ACTA with the aim of protecting intellectual property and copyright. The Senate's Permanent Commission refused to ratify the law on the grounds that it violated already established laws regarding international economic treaties.

---

14   Senado de la República. Gaceta del Senado, "Reforma a la Ley Federal del Derecho de Autor y adición a la Ley de Propiedad Industrial", [Accessed 25 May 2014]. Available at: http://www.senado.gob.mx/index.php?ver=sp&mn=2&sm=2&id=12788&lg=61.

15   Anonymous Iberoamérica. "Blog Oficial, #OpDoging", Blog Oficial, [Accessed 2 February 2015]. Available at: http://anonopsibero.blogspot.com/2012/01/opdoring.html.

Towards the end of March 2014, President Enrique Peña Nieto sent an initiative to the Mexican Senate designed to expedite secondary legislation regulating the telecommunications and broadcasting industries. In section V of Article 145, companies were authorized to violate the neutrality of the Internet, while Article 197 in section VII gave the government authority to "block, inhibit or temporarily cancel telecommunications signals in critical events and places in the interests of public and national security when acting at the request of the competent authorities." In addition, the initiative required Internet servers to store their data on users for up to two years and to make them available to state authorities upon request. This proposal known as the Telecom Act came to be challenged by activists, civil society organizations, academics and thousands of Internet users. Although the Telecom Act was passed on 8 July 2014, the final version of the law was watered down with the elimination of the controversial articles concerning the loss of neutrality and the blocking of telecommunication signals.

Conclusion

In the global world of the Twenty-First Century, the Internet has become perhaps the most open form of mass media for the exercise of human communicative freedom. The greater degree of autonomy that cyber networks offers its users, the possibilities of sharing content through a free and open culture, and an underlying collaborative philosophy, can turn the elements of informational instruments into means of social progress and of expanding the possible forms of power by establishing greater organizational links, all of which favours a greater capacity of mobilization with others.

But such freedom is constantly faced with enemies such as governments and powerful groups that seek to establish limits on the ways in which people can connect up with their peers. In recent years, a series of measures have been implemented around the world that would seek to erect barriers around the free flow of information through the implementation of Tools of Coercion (TC). These coercive instruments are used as desired by visible or invisible public policies and create broad legal frameworks that seek to criminalize those cyber activities considered a threat. Under the pretexts of fighting terrorism, drug trafficking and piracy, states have waged a campaign against the "evil" that is lurking amidst civil society. Global online espionage, the complacent participation of companies in handing private information over to law enforcement bodies, and the mechanisms of domestic surveillance that have been discovered so far (along with those which remain hidden from view) all aim in the final analysis to control parts of the vast open space constituted by the Internet. This form of coercive penetration seeks not only to enter into the private domains of users but beyond that to modify the circuits of information stored in the very minds and ideas of people.

CHAPTER 5

Grey Areas in China's Growth: A Questionable Development

Silvana Andrea Figueroa Delgado

Since ancient times, China has distinguished itself as a great nation in many ways. Its contributions to humanity are undeniable. It was the birthplace of printing, the magnetic compass, silk, the first earthquake detector, the mechanical clock and non-submersible vessels, among other crucial inventions. It is also said that China had a high level of literacy from very early times, especially with respect to philosophical and spiritual matters (See Oubiña Falcon 2015 and Pato 2011).

Just as in earlier times, the People's Republic of China today sparks great interest for scholars of economics and related issues. China's major contribution to the world's gross domestic product (GDP), according to the United Nations Educational, Scientific and Cultural Organization (UNESCO), amounted to 10.7 % of the global economy in 2007. China conducts the largest percentage of global trade and receives the most significant level of direct foreign investment, all of which figure as elements that draw attention to this country (UNESCO 2010).

Our intention in this work will consist in identifying the key elements of this spectacular economic growth that China has experienced over recent decades. We will begin by presenting our notion of development and drawing out the variables that implicitly underlie the development process. That will lead us to consider the strategies adopted by China in order to fortify its scientific and technological development, strategies that have reached a critical mass. This approach allows us to analyse both the achievements as well as the social costs incurred. By beginning with this comprehensive perspective, we will then be better able to consider the political order and the state of civil liberties, being that such elements form inseparable parts of the model in place. In this manner, it will be possible to offer important points of debate about the existence of real democracy. At the end, we will offer some reflections that seek to interrelate these monumental elements of economic growth, development and democracy.

Advances and Impediments to Development

We have understood capitalist development as a qualitative condition, reflected in the capacity to create technological progress, transforming innovation into a constant process that is tightly tied to productive processes where its general diffusion allows for the homogenization of the economic structure (Figueroa Delgado 2008).

This form of development presupposes national independence that is granted through a solidly endogenous scientific and technological platform. The presence of these factors permits a base for social welfare that should be reflected in rising wages, itself a product of the greater skill levels it requires of the labour force. These factors also provide for expansion of occupations tied to the opening of new sources of production—sustained by constant innovation—all of which should result in a more equitable distribution of income and a substantial reduction in poverty. Other social responsibilities are susceptible to increases of state fiscal growth derived from the steady overall growth of economic activities. Scientific progress should become crystalized in the general increase in quality of life for the entire population in view of the expanded potential to attend to health care and the environment.

This essential condition thus described as a genuine capacity to develop science and technology is indeed a central condition for enabling a nation to aspire to achieve development. In this area, China has demonstrated significant advances. For 2011, it accounted for 15% of global investment in research and development, placing second worldwide. Towards the interior, research and development (R&D) represented only 1.84% of GDP given the tremendous amplitude of the latter (National Science Board 2014). The present public policy for science emerged in the decade of the 1980s in the context of larger reforms of the economic system towards greater opening.

Previously, the scientific and technological efforts had privileged areas of national security such as "nuclear weapons, outer space and the synthesis of insulin" (OCDE 2008: 384). This took place under a scheme of state centralization that was highly influenced by the Soviet model. The economic structure was dominated by strategic state enterprise monopolies and the principal research agents were "the Chinese Academy of Sciences (CAS), the ministry-affiliated academies, the R&D institutions affiliated to provincial governments, the universities, and defence" (OCDE 2008: 383–384).

With the economic liberalization, the aforementioned scheme became modified. Ideologically, the new premise that now circulated was that with the participation of private enterprise, technological progress would be better focused on development while the actors would further multiply, creating tighter links between the creation and utilization of knowledge.

As the OCDE *Reviews of Innovation Policy* (OCDE 2008) shows us, China moved to create the institutional, economical and organizational climate that it considered propitious for technological development, while it committed a whole array of ministries and state agencies to the task of building up the scientific-technological platform. A whole series of laws and mandates were promulgated in which public policy measures were put into effect that regulated investment, fiscal incentives, financial support or subsidies, state demand and risk capital.

Among key decisions taken was the reduction in the financing of operational costs of the public research centres so as to obligate the search for complementary funding and enable sale of their results. Also, the establishment of the National Sciences Foundation and funding for risk capital, and the prioritization and concentration of resources into larger programs such as the National High-Tech Research and Development Program (863 Program), among others.

China established important special zones to house companies committed to the creation of high technology. These included a support package that offered favourable commercial treatment, the installation of technology transfer offices and centres of technological promotion, along with a whole repertoire of support aimed to engender the training of staff in research and development.

This vast collection of measures produced valuable results: an expansion of the entrepreneurial sector and R&D laboratories; a greater linkage between the institutions of higher education and the private sector; a very active export market, including products of high technological content; a significant growth in patents and published articles being cited; in addition to becoming the nation with the largest number of full time researchers after the United States (UNESCO 2010). It is worth noting that since 2009, the domestic patents granted for inventions surpassed those being granted to non-residents and since that time, the gap has continually widened (NBS 2011; 2014).

All of this notwithstanding, there also exist signs alerting us to fault lines or obstacles in the process and which call the Chinese path into question. If we focus in on the patents being granted, we can see that even though Chinese researchers have gained the upper hand on their foreign peers, the portion dedicated to inventions remains small in relative terms. Only 11.68% of the patents granted to residents pertain to inventions while 75.84% of those awarded to foreigners fall into that category according to 2013 data (NBS 2014).

In a brilliant work written by Yuqing Xing (2012), the true significance of China's ascent as the world's number one exporter of high technology goods is debated. The author discussed two key aspects regarding goods that domestically represented a third of total manufactured exports in 2010. First, 79.9 % of these goods being sold abroad fall into the category of "processing trade," i.e., they correspond to product that are processed and assembled from imported materials, parts, and components, either partially or completely, before being resold on the global market. He argues that in these cases, the real aggregate value is given by Chinese labour, rather than the technology being applied, and that their denomination should be changed to "high-tech assembled goods." Moreover, he demonstrates based on studies by Xing and Detert (2010) and Dedrick, Kraemer and Linden (2010) the essential smallness of value being added (Xing and Detert 2010: 81–116). Secondly, 67 % of these exported goods belong to firms that are essentially foreign owned, and 82 % of the companies share significant foreign capital participation, such that only 17 % of the firms are local in nature. All of this is the inheritance of the broad economic opening that China had embarked upon.

The Chinese liberalization, as we mentioned, has its origins in the decade of the 1980s and was deepened in the 1990s when China formally adopted the model of "market socialism", a term that alluded to the combination of a planned economy with a market economy, placing greater importance on the latter in the formula (Peiyan 2012). Direct foreign investment was permitted throughout the country, no longer just in specific zones. While small and medium sized state companies became Limited Liability Companies, the large state firms became Publicly Traded Companies. The state kept the most strategic enterprises in its own hands (in, for example, the areas of electricity, petroleum, steel, telecommunications, banking, mining, etc.) (OCDE 2018: 140). In 2001, China joined the WTO.

Against this backdrop, it is undeniable that China realized a tremendous effort in promoting research and development. Yet, it is difficult to speak of economic sovereignty while one's technological horizon is so impacted by transnational corporations. Another negative element implicit in China's growth would appear to be the prevalence of low wages, which is precisely what attracts foreign investment, land expropriations and ecological contamination.

If indeed the 2012/13 Global Wage Report of the International Labour Organization (ILO) views with optimism the wage growth that China has experienced over the 2000–2010 period, upwards to double digit annual average growth, the fact is that it even remains low. "For example, in 2010, the average monthly wages in the United States were around US$3,300. In China, wages ranged between $250 in the private sector to US$440 in the public enterprises. If these figures were adjusted in order to consider the lower cost of living in China, they would range between $400 and $700 monthly income" (ILO 2012).

Nor should we be remiss in recalling the many documented cases of extremely poor working conditions for Chinese labourers.[1]

As was already mentioned regarding human resources employed in the field of research, China figures in second place globally, but not when adjusted with respect to its large population, i.e., per thousand of economically active persons. In fact, if adjusted in this manner, China is far out of the highest global tiers.[2] This is of course due to the fact of its immense population, which accounted for almost 20% of all humanity in 2014 (Worldometers 2015), and also due to the prevalence of an extensive rural citizenry that has made it difficult to extend higher education at the optimal pace desired.

The rate of urban unemployment was 4.1% in 2013 (EFE 2014). However, there was no official measurement for the rural situation. It was estimated that 128 million Chinese lived in poverty in 2011, which represented 13.4% of the population (CIA 2014). Rural and urban lands are expropriated forcibly in China with little margin for opposition, although reports do exist of collective peasant and community confrontations with the police. Expropriated lands according to *The Epoch Times*[3] are compensated for, but not always in a comprehensive manner on account of corruption on the part of government officials. These lands are destined for the installation of industries, housing developments and even recreational facilities such as golf courses.

The rush to attract foreign investment has been to the detriment of the natural environment. There are various studies and documentaries on what are now called "cancer villages," located mainly in rural areas that have hosted chemically toxic industries, which have badly contaminated the air, water, food and have affected the surrounding population (McKenzie 2013). In coastal areas, the dumping of waste at sea has been documented (Guang 2009). Carbon emissions qualify as among the highest in the world: 7 metric tons per year per inhabitant (APF 2013). While the per capita figure suggests that there are many countries with a higher amount, the absolute magnitude is enormous.

The costs of China's growth have been high and varied. In order to present a more complete analysis, we now turn to the political realm and to the rights of Chinese citizens.

---

1   For example, see Redacción Web (2014).
2   See UIS (2015).
3   See various articles under "Expropriations in China" as available at: http://www.lagranepoca .com/archivo/category/free-tagging/expropiaciones-en-china.html.

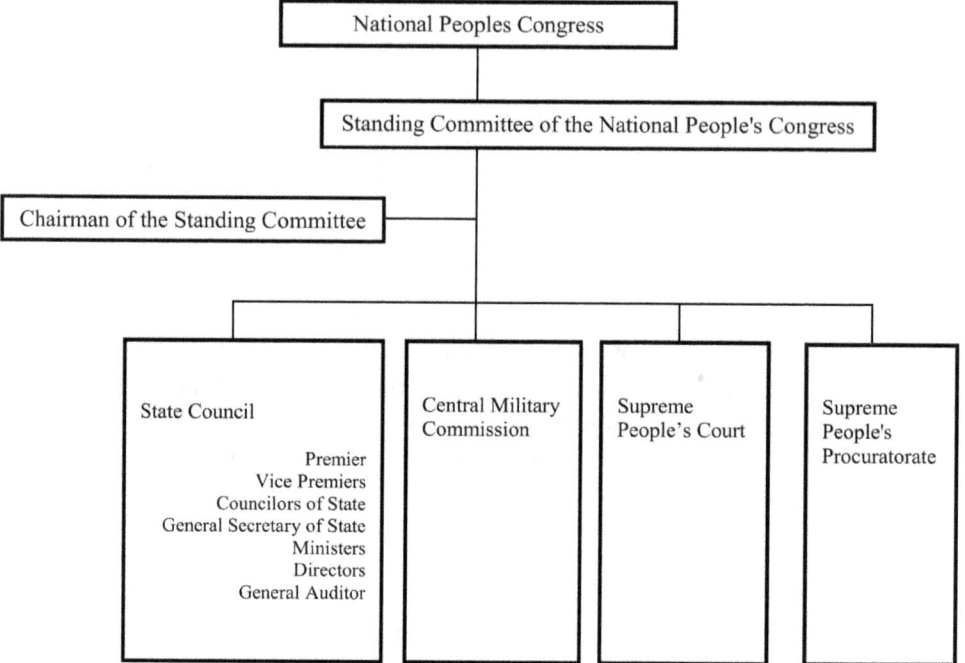

FIGURE 5.1 *Diagram of the National People's Congress*
SOURCE: PEOPLE'S DAILY, COUNCIL OF STATE, ORGANIZATIONAL
STRUCTURE OF THE NATIONAL PEOPLE'S CONGRESS (MOFCOM 2012B)

Democracy Issues

In the area of political institutions and state power, China can be seen to have a complex legislative structure. The People's National Congress (Figure 5.1) formally constitutes the highest legislative authority. It is made up of around 3000 deputies,[4] of whom a little more than 5% or 150 constitute the Permanent Committee (Martínez 2012). Among its diverse functions are to elect the President of the People's Republic of China and its Prime Minister. The latter heads the Council of State which is made up of the vice-premiers, councillors of state, ministers, commission directors, general auditor, and the secretary general (CRI 2013) of state, all of whom are designated by the Assembly to which it answers (MOFCOM 2012b).

---

4 Candidates may be nominated jointly or independently by political parties, mass organizations or more than ten voters, but the number of candidates a voter nominates shall not exceed the number of deputies in the corresponding areas (MOFCOM 2012a). According

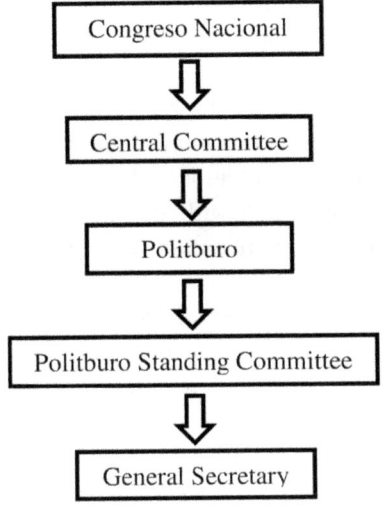

FIGURE 5.2
Simplified diagram of the Communist Party of China
SOURCE: CONSTRUCTED FROM INFO
CONTAINED IN MOFCOM (2012D)

The State Council (Figure 5.2) has the responsibility for implementing laws and decisions of both the National People's Congress and the Communist Party of China (CPC) (Diario del Pueblo 2000). Contrary to what happens in so-called "modern democracies," China's Constitution recognizes the CPC as the sole ruling party (MOFCOM 2012c). Decisions issued by the Central Committee or the Politburo Standing Committee (PSC), appointed by the former, have the greatest influence on the direction of state policy in any area of public interest.

The structure of the Party hierarchy includes the CPC National Congress that elects the members of the Central Committee, and the latter is formally accountable to the former. But in turn, the Central Committee decides on the number and election of delegates to the National Congress. Similarly, the Central Committee designates the Politburo (members with positions in the Council of State) and its Standing Commission, consisting of a smaller select number who in turn elects the Secretary General of the CPC (MOFCOM 2012d)

---

to Martínez (2012), the electoral system operates at various levels: 1) the People's National Congress; 2) the Provincial, Autonomous Region, and Municipal Popular Assemblies under the Central Government; 3) the Popular Assemblies of cities divided into districts and autonomous prefectures; 4) the Popular Assemblies not divided into districts, municipal districts, counties or autonomous counties; and 5) the Popular Assemblies of municipals, municipals of ethnic minorities and peoples. First, the community designates their local representatives, then, these elect those who will represent at the provincial or regional level, and the latter of these select the candidates who will represent as deputies to the National Popular Congress.

who likewise serves as the present day President of the People's Republic of China and Chairman of the Central Military Commission (Abahaj 2015).

In addition to the hegemonic party, eight other political parties exist in China.[5] These other "contestants" of the electoral system are recognized not as competition to the CPC but as complementary to the work of the Communist Party, sharing close ties of collaboration. Some members of these parties are invited to the congresses of the CPC, although without vote, or to other symposia or discussion forums where they can express their views (MOFCOM 2012e).

This feeds the notion of "democratic centralism" as elevated to a constitutional level (MOFCOM 2012f) and this organizational principle also guides the CPC (MOFCOM 2012d). It translates into decisions made by a few which are then ratified by others. This likewise applies to the coordination between the central government and local authorities, the latter of which operate with a measure of autonomy to receive and adapt national laws to local regulations and circumstances, particularly given the multiplicity of ethnic groups and regions that can hardly be treated in detail by the general laws.[6]

What has been said so far speaks more of a political system with strong authoritarian features. But there are additional elements that question the very existence of genuine democracy in China. Amnesty International has documented cases that suggest a systematic violation of human rights in the nation (Amnesty International 2015). Torture and death sentences rise to alarming rates, without publication of any official records, and this has been questioned for some time by AI which estimates thousands of victims and alleges that the number of death sentences in China exceeds those being carried out in rest of the entire world. A major problem is that capital punishment can be applied for non-violent crimes and for those confessed to under torture. The mere fact of questioning the regime can earn a judicial rebuke, thus making it a very dangerous practice, even for lawyers engaged in the defence of their clients.

There have been well known acts of censorship carried out by the Chinese government. In April 2013 following the award of a Pulitzer prize for an article published in *The New York Times* regarding the riches acquired by former

---

5   They consist of the Revolutionary Committee of the Chinese Kuomintang (RCCK); The China Democratic League (CDL); The China Democratic National Construction Association (CDNCA); The China Association for Promoting Democracy (CAPD); The Chinese Peasants and Workers Democratic Party (CPWDP); The China Zhi Gong Party (CZGP); The Jiu San Society (JSS); The Taiwan Democratic Self-Government League (TDSGL) (MOFCOM 2012e).

6   There are also "autonomous regions and that of the special economic zones and the Hong Kong and the Macao Special Administrative Regions" (MOFCOM 2012b).

Premier Wen Jiabao, the state authorities decided to ban any publication of correspondents or foreign media, as well as those of independent journalists, NGOs or any commercial entity otherwise not verified by the state. Similarly, the state warned those responsible for the administration of local networks to refrain from encouraging rumours that harm national interests (EFE 2013).

The media censorship also extended to movies and television programs being transmitted by social media networks, citing "political reasons." In April 2014, the programs "The Big Bang Theory," "The Good Wife," "The Practice" and "NCIS" were all retired from broadcast (EFE 2014a), suggesting considerable sensitivity to criticism and denial on the part of the Chinese government. Moreover, it seems that in terms of diverse religious beliefs and practices, there are some that are tolerated, but not all. Particularly in Tibet, constant repression of peaceful demonstrations that manifest ethnic, religious and cultural identity has been reported (Amnesty International 2014).

Despite a context of repressive authoritarianism, demonstrations do continue to be reported notwithstanding the risks involved with challenging the state. A report published in *El País* estimated that there were around 180,000 "protests, strikes and incidents of social unrest" throughout the country in 2010 related to environmental, labour, or land expropriation issues, as well as government corruption. More recently, protests regarding the freedom of expression have also been reported (Reinoso 2013).

Conclusion

China has laid firm foundations for the deployment of scientific and technological capabilities. This explains to a considerable extent why the country has been a preferred international destination for foreign direct investment. With its cheap labour and attractive tax incentives along with other facilities for the installation of new investments, China has characteristics that stand out among nations competing for direct foreign capital. Therefore we can safely reaffirm the advantageous status of China in this regard. We will have to recognize that the high quality and skill levels of its intellectual and manual labour force have dovetailed with a sophisticated and accumulating R&D infrastructure backed by a package of state supports for the promotion of scientific and technological processes. In another study, we showed that it is the very laboratories of R&D that initiate a practice of international diffusion, so even if the nation of China does have the highest percentage of these technological generators, it does display one of the highest growth rates in their multiplication.

According to the 2014 China Statistical Yearbook,[7] total investment as seen by the expenditure side (expressed as gross capital formation) accounted for 47.7% of GDP in 2012. While data from the United Nations Conference on Trade and Development (UNCTAD 2015a) indicates that of this investment only 3.14% were made up of foreign inflows for that year, the fact is that important economic sectors of the Chinese economy are led by foreign capital. Other data indicate that in 2013, foreign companies were responsible for 47.5% of Chinese exports, especially concentrated in the area of high-tech products (Krokou 2015). It was already the case by 2010 that machinery and electronics exports accounted for almost 70% of the activities of resident foreign firms (Agencia de Noticias Xinhua 2010). When in 2014 China became the nation that captured the world's largest share of Foreign Direct Investment (UNCTAD 2015b), these percentages should not be all that surprising. It should be further noted that the outflow of Chinese capital has also expanded and the fact that the preferred sector has investment in services such as manufacturing accounts for only about 23% of those outflows (Krokou 2015).

In this way, the impressive scientific-technological effort of China has been the object of considerable exploitation by external agents aiming to satisfy their own interests. Given its broad level of economic liberalization, the country has not successfully achieved the most optimal use of its resources. Transfer is notorious since domestic firms do not lead in the manufacture of high-tech goods and sometimes the potential of skilled labour is not fully exploited by those Chinese industries largely confined to assembly tasks.

To this, we have to add the fact that China has relegated considerable portions of its territory to highly polluting activities to the detriment of its environmental and human health profiles, which in many cases signified further expropriating peasants or urban families from their homes. Nor has the treatment of workers always been the best, as wages have remained relatively low, even after the increases of recent times, while in some cases unacceptably poor working conditions persist.

State violence is visible in various areas, and sometimes in a very direct way such as: 1) its oversight of important social events; 2) in the area of censorship, and 3) in the form of violations of human rights. This has amounted to a political system with little margin for effective participation, with abuses of power that extent all the way to torture and excessive application of the death sentence. It can certainly be argued from a liberal standpoint that democracy

---

7   See Table 3–18 NBS 2014.

in China is absent, even in its most basic formulations of civil and political liberties, despite the fact that these rights are printed in the Constitution (MOFCOM 2012c).

In short, although China displays an understanding of the essence of development, its full consolidation has proved elusive on account of the tremendous economic opening it has relied upon. In other words, its ascension has been truncated despite its multiple efforts. We could even say that the form of growth China has adopted has itself constituted the drag on its development and its democracy, and that this reveals an absence of a truly national project, given that local actors can be the first to reap its benefits.

CHAPTER 6

Economic Growth, Democracy and the Construction of Citizenship in South Korea

Cristina Recéndez Guerrero

Introduction

This study explores the factors that promoted, paced and sustained the economic growth and development as well as the political system of South Korea (the Republic of Korea). Organized into four parts, the first section addresses conceptual aspects beginning with the category of "bureaucratic authoritarianism" (O'Donnell 1982) as a category that helps us define the political and democratic features within a given social order. In the second part, the general background of economic and political growth of this nation (1960–1992) is surveyed, addressing the genesis and consolidation of the bureaucratic state of South Korea as based in the era of military governments. The long period of obstacles and hardships that produced struggles and movements of political actors aiming to achieve the enjoyment of citizenship is seen as paving the way to demands for changes in the political system that ultimately would move the country towards democratization.

The third part of this work discusses the factors that led to the society becoming more politicized (1993–2002), initiating the process by which social movements and protest ended in negotiations that brought an end to the "bureaucratic authoritarian" political system and opened the way for civilian representatives to exercise governance. Emphasis is placed on the Asian financial crisis and its effects upon economic growth and the formation of political parties. Finally, the fourth section discusses some of the particular characteristics experienced during democratic normalization (2002–2013) in a context of sustained economic growth.

Conceptual Aspects

Political Science has entertained a longstanding debate on how to best study power and the state. This debate has raged because the distribution of power is at the centre of democracy in which the state serves the function of ensuring to all individuals the full exercise of their rights. However, democratic transformations do not always occur at the same pace as economic growth, that latter of which in South Korea generated high levels of social differentiation. To analyse the economic growth and democratic development of South Korea, we will base our consideration upon the following definition: "democracy is defined as a political system with a separation of powers, guarantees for individual rights, and governing bodies of popular representation. In turn, the political system is defined as the set of institutions, organizations and political processes, characterized by a certain degree of interdependence which govern and shape the political life of a particular community" (Aguirre 2009: 10–11).

In a political system, it is the society that places a particular person in a special social stratum, delegating to him or to her concentrated use of power. Nevertheless, it does not always work that way. There are special interest groups and external relations of force that, through the exercise of symbolic relations, manage and consolidate the basic suppositions of economic policy. A political system can over the decades lead the development of a country through actual subjugation of the population.

In this sense, Guillermo O'Donnell considers "domination (power) as the current or potential capacity, that regularly imposes its will upon others, including although not necessarily against their resistance" (1982: 62). With regard to the resources of power, he suggests that they can be broken down into physical coercion, ideological control, control of economic resources and control over informational resources. The control over any of these allows for coercion, where the dominated subjects accept as fair and natural the unequal relationship that they are party to and thus submit to.

In that light, the analysis of distinct relations of force and the exercise of power that ruled the economic development of South Korea fits the profile of O'Donnell's (1997) "bureaucratic-authoritarian" system.[1] O'Donnell's model exhibits the general features of political exclusion: scant democracy; dominant actors represented in the bureaucracy and technocracy, civilian and military, collaborating with foreign capital; absence of free electoral competition; strict control exerted over political participation of the various social classes; and the

---

1   His model was presented as a conceptual frame of reference that serves as a point of departure for explaining when there are various interactions between political factors and social and economic change, i.e., it combines an analysis of the structure of economic development with the political regime (freedom of electoral competition, interest groups and the level of political repression), with the interests of who governs (class and sectoral composition of the dominant political coalition), and who benefits (distribution of resources through public policies) (O'Donnell 1997).

promotion of public policies aimed at consolidating an advanced industrialization, promoting the process of capital accumulation for the benefit of private capital.

According to O'Donnell and Philippe Schmitter (1991), two moments can be identified within the system of "bureaucratic authoritarianism" in the process of building democracy. First, a transitional stage that begins with the decomposition of the authoritarian regime up through the installation of a democratic government, including the celebration of free elections and the orderly transfer of power to the elected government. Second, there is a stage of consolidation when the system becomes converted into a democratic regime characterized by solid, strong and consolidated institutions and decision-making spaces at the core of the political system. It further suggests that the transition can occur in different ways. In general, it could happen from above, agreed upon and controlled by the outgoing regime, or, it can emerge from below, involving a rupture that is initiated by the actions taken by opposition forces, either by bringing down the government, or by other means forcing the preceding regime to retire. In the case of South Korea, the change emerged from below.

With respect to the economic structure, O'Donnell (1982: 15–17) suggests that what characterizes a capitalist society per se are its relations of production, and in this economic formation exist relationships of domination and subjugation, i.e., capitalists and workers. These relations are uneven and contradictory, and are set throughout the process and workplaces. The state forms part of these social relations, ensuring the coercive relationship that capital requires for its operation and reproduction. At the same time, it is the state that organizes capitalist social relations and the domination that is required for the existence and reproduction of social classes, i.e., dominated classes. At the concrete level, the state, the bourgeoisie and the workers are objectified in actors or social subjects and institutions, or the state apparatus. However, the state-apparatus, just like commodities, is fetishized and appears as an external third party to the subjects of social relations, rather than appearing as part of them. This appearance of externality enables the state to be constituted as organizer of bourgeois domination.

Alfredo Romero (2001; 2012) argues that a model of economic development based on two pillars was created in South Korea, namely, the state and "*Chaebol*" or industrial enterprises or conglomerates under state direction (Grou 1988: 42–43).[2] The *Chaebol* were created in order to develop an export economy

---

2   The *Chaebol* consists of private companies whose capital is financed by the Nationalized Bank with direct participation of state investment, low interest rates, a low level of stock dispersion,

over the capital-labour base where the role of the bureaucratic-authoritarian state was controlling the pace and goals of production, the terms of trade, and the consciousness of workers so as to inculcate within them the goals set for production. As other authors also argue, the authoritarian developmentalist state maintained under its control the base of the capitalist-labour relations of production.

Nelia Bojórquez cites T.H. Marshall in asserting that citizenship can be conceptualized as "a process of social construction linked to the dynamics of democratization which together integrate civil, political and social rights as a social status that determines the sense of belonging in a national community and encourages participation in social life" (2005: 1). Citizenship is the effective exercise of rights and if any of these do not exist or if they are not respected, citizenship is practically non-existent.

Following this argument, civil citizenship confers the rights of individuals before the law and guarantees them the right to live according to their own choice, having personal freedom and freedom of beliefs, and promotes their rights to property and justice. Political citizenship in turn confers the right to elect and to be elected, social rights and the right of every person to have at least a minimum level of economic well-being and security (i.e., the rights to social security, wages, and social welfare among others).

Alicia Iriarte, et al. echo O'Donnell by stating: "historically, citizenship arose along with capitalism ... the citizen is who has the right to perform acts that result in the formation of the power of state institutions, in the choice of leaders who can mobilize those resources and demand obedience, and the ability to resort to pre-established legal procedures for protection from arbitrary intromission" (2003: 16–17). In the context of the development and economic growth of South Korea, the present work argues that the political, civil and social rights that make up citizenship were achieved in a little over three decades; the agent for obtaining the recognition of citizenship was the social movements of students and workers who in the face of relentless injustices convinced the majority of the population to participate in the democratization process.

---

with effective control over the company in the hands of few individuals or families, and initially, formed with nearly a total absence of foreign capital. Examples are Samsung, LG, Daewoo, and Lucky Goldstar. See Bustelo (1991).

South Korea's Acquisition of Bureaucratic Authoritarianism

In light of the context established above, we now turn to the socio-demographic and economic transformations that have taken place in the over 60 years of the Republic of Korea, located south of the 38th parallel on the peninsula. The nation's parliamentary democracy and presidentialist government[3] rules over an area of 99,900 square kilometres. Its population of approx. 50,236,000 people exhibits a natural growth of 0.5 %, a population density of 512 inhabitants per square kilometre, with 83.5 % of the population living in urban areas (16.5 % rural) and 49.7 % male (50.3 % female). It is the thirteenth largest world economy (13/196) and ranks 31 among the countries with the highest GDP per capita ($ 33,200 in 2013), with a Human Development Index (HDI) in 2013 of 0.891 / 1 that put it in 12th place. Its economic model is based on exports that account for over 40 % of GDP, making Korea one of the most successful cases of state economic development planning under a purely capitalist system. It is among those countries that most recently entered into the OECD (1996) and in 2007, it joined the Economic Commission for Latin America (CEPAL).

As Jose Luis Leon (2006) argues, South Korea was at a serious disadvantage for achieving economic development, democracy and citizenship for its population. First, it had a monarchical government imposed by the Japanese occupation (1910–1945) prior to the division of the peninsula. This monarchy inherited a classic kind of East Asian authoritarianism and state control that was honed over the centuries. Second, it possessed a backward industrial base and a financial system that had been in the hands of the Japanese. Third, the bloody internal war that took place between 1950 and 1953 ended with the economic, political and social division of its national territory creating North Korea (as a Russian "protectorate") and South Korea (a "protectorate" of the United States). All of this notwithstanding, South Korea emerged since the last quarter of the Twentieth Century as one of those South-eastern Asian countries that went from a poor, rural economic regime based on farming to a highly industrialized economic regime whose per capita GDP grew at a 6 % annual

---

3   Executive power is exercised by the President who names the Prime Minister with approval
    of the Parliament. The President is the head of state, chief executive of the government,
    and commander in chief of the armed forces. The President names 17 Ministers who receive
    advice from 7 independent counsels, directs the National Intelligence Agency (NIS), and
    the Prime Minister assists the President in supervising the Ministers, and acts as interim
    President if the need arises.

average, with a high degree of urbanization, and a growth strategy based on exports and information technology driven development.[4]

There are several studies[5] that describe how South Korea's production relations were handled under authoritarian and militarized direction. They review how capital accumulation came to be pervaded by relentlessly conflictual relations between the state, employers, trade unions and citizens. Jorge Santarrosa (2005) describes how its growth was shown to be engineered through heavy state intervention that planned the industrialization process and how two social classes became consolidated during this phase, the bureaucracy and the technocracy who together were responsible for a well-managed, planned economic development for the benefit of capital. It's worth noting, however, that the process did not pave the way for recognition or transformation of any other social classes that formed part of the system.

On 15 August 1948, the Republic of Korea was proclaimed by its first President, Syngman Rhee (1948–1960). Rhee was supported in the presidential elections by what was to later become the Liberal Party and was voted in by direct popular election under the supervision of military troops and UN observers (Aguirre 2009). Rhee's First Republic presided over an emergent shift towards authoritarian power and abuses against the population designed to maintain the regime in power. As he neared the end of his first four-year term, his regime introduced martial law in May 1952. In 1955, the National Assembly[6] was forced to make an amendment to the Constitution that granted Rhee the prospect of continuous re-election as a wave of repression ensued, including outlawing The Progressive Party, a moderately socialist opposition party.

As Leon (2006) argues, Syngman Rhee maintained rule through the support of the national army and US military forces that anchored his personalistic rule for over a 12-year period. During that time, he was re-elected in 1956 and 1960 by a large majority through elections marked on both occasions by widespread fraud. Opposition was expressed among student organizations who joined in common cause to protest each re-election and this helped create the nationwide student movement against arbitrary rule and the lack of democracy.

---

4   South Korea forms part of the four south-eastern Asian countries that constitute a special case of economic development along with Taiwan, Singapore and Hong Kong.

5   See León (2006); Romero (2003); Silvert, Santarrosa and Bauer (1997); and Figueroa (2012).

6   Legislative power in South Korea is composed of a unicameral parliament known as the National Assembly that has 299 members elected to four-year terms, and normally convenes 100 sessions per year. See: http://observatorio.bcn.cl/asiapacifico/noticias/sistema-politico -corea-del-sur [Accessed 22 March 2014].

Organized students and the middle classes ultimately emerged as the fundamental force in the overthrow of the dictatorial Rhee Regime. The conflict between the government and the population led to what became known as "the Revolution of April 1960" which resulted in the collapse of the form of political democracy that had been installed by the US. Carlos Penoncello (2005) asserts that this first student movement was important because it constituted the first stage of the struggle for democratization. It was also during this period that the country's economic growth was based on its offering cheap labour for foreign industry, consistent with the historic role perpetrated by Japan and the US.

Following the overthrow of the Rhee Regime, an amendment was made to the Constitution, and the Democratic Party elected Yoon Poh as President and Chang Myon as Prime Minister. With this election, the path was opened to a parliamentary system and the restoration of political citizenship (Aguirre 2009). However, the fledgling democracy suffered a major setback in 1961 when Park Chung Hee led a military coup and ousted the president, suspended the constitution, banned all political parties, and dissolved the National Assembly through the imposition of martial law. With these actions, the political regime of "Bureaucratic Authoritarianism" initiated earlier by Rhee was re-imposed and consolidated, led by an army that proved to have little reluctance to resort to atrocities and numerous forms of injustice against the population.

Park Chung Hee emerged from the coup as the President of the Second Republic of Korea (1961–1979). As Han Sang Jin (1997) indicates, it was the student movement that reinitiated the struggle for democracy under this dictatorial regime, openly posturing as opponents of the regime and leading a struggle around three main lines: 1) democratic orientation, 2) orientation towards equity, and 3) nationalist orientation, demanding national self-determination and the reunification of both Koreas. Park's authoritarian rule was legitimized by the "elections" of 1963. Supported by the military from where he emerged, Park launched his system of economic planning with its primary goal to achieve industrial development and open the market to exports. In the context of the Cold War, the Park Regime received financial aid from the United States that totalled $3.1 million and had the firm support of those multinational corporations present in the country (Toussaint 2006). During this period, the United States pressured the South Korean government to resume relations with Japan, a reconciliation that brought $500 million in aid to the country, payable over ten years (1965–1975), of which $300 million was granted as a donation.

With Park in power, the priority goals were economic growth based on industrialization and maintaining his political control. The development of planned industries was launched with mechanisms in place at the onset for state intervention in order to protect and participate in the process. With national and foreign capital, the largest *Chaebol* were formed, the first generation of which were oriented to the development of light, labour intensive industries such as textiles, garments, electronics assembly and foodstuffs. The manufacturing sector grew at an 18% per annum rate, and during the first three five year plans (1961 to 1976) the national economy as a whole grew at an average annual rate of more than 10% and real per capita income tripled (Toussaint 2006).

Various studies have revealed that the state and *Chaebol* companies were the main political and economic actors of South Korea, two institutions that appropriated all power and formed a very strong nucleus that together closed off all possibility of any opening to a democratization of the country. For these institutions, democracy represented an attack on productivity, industrial discipline and the requisite strict control over workers. In spite of the increasing size of the working class, the formation of independent trade unions remained prohibited.

In 1962, the path into the Third Republic would be initiated with an Economic Development Plan for the following five years already approved. The strategy was the implementation of policies of selected promotion, including among its objectives the development of heavy industry and the creation of a higher quality system of education in support of growth in scientific and technological fields. Julio Rubio, et al. (2013) argue that the promotion of research and development was supported, regulated and protected by a system of laws created according to the needs of development. The first law issued by the government had the development of atomic energy (1959) as its main concern and was enacted to promote the peaceful use of nuclear fission and the establishment of preventive measures and public safety in case of nuclear disaster.

The industrialization project required all producers and employees to fight for capitalist economic development, and for this Park exhorted a nationalistic spirit. However, the right of citizens for a more equitable distribution of income was not part of the formula. Instead, Park sought to maintain the discipline of workers and increase his political legitimacy in the face of ignorance and illiteracy in the rural areas where little education, medical and social services could be found. In searching for better job opportunities, young people immigrated to the cities and swelled the ranks of unskilled workers, leading to the emergence of urban shantytowns (Díaz 2005).

During the Park regime, some institutions of social security were created, initially only for civil servants (1961) and military personnel (1963). Public assistance and social welfare services for people with special problems was created in 1961 and an accident insurance plan was created in 1964. Despite a social security law being enacted in 1963, nationwide health care would not be implemented in South Korea until 1977 (Valencia 2001: 103–104). All of these institutions constituted attempts of the Park regime to garner international legitimacy.

Another area in this regard was education. In order to advance economic growth and industrialization, the development of science became viewed as a priority. The state promulgated and has continued to update laws in this area. In 1967, the Law on Promotion of Science and Technology (Law No. 1864) established the legal basis for the construction of the national scientific system and the creation of the Ministry of Science and Technology (MOST). This was later changed to the Ministry of Education, Science and Technology (MOEST) during the Bak Lee Myeong Administration. According to Rubio, et al. (2013: 32) the 1972 Law of Technological Development (Law No. 2399) established the fiscal and financial incentives needed to stimulate investment in research by the private sector.

The same approach was used to promote industrial engineering, manufacturing and commercialization of the results of research and development when the MOST created the Law for the Promotion of Engineering in the Area of Services (Law No. 2474 of 1973). Subsequent laws in this vein included the Law on the Promotion of R & D (1972); the Law for the Promotion of Pure Scientific Research (1989); the Law on the Promotion of Small and Medium Enterprises (1995); the Special Law on Scientific and Technological Innovation (1997); the Law on Technology Transfer (2000) and the Science and Technology Act (2001). The MOEST is the state organ that exercises control over the 1,134 existing laws that govern South Korea's policy of promoting scientific application through technological development, human resource training and the development and peaceful use of nuclear energy.

In the context of a political reform, Park announced the *Yushin* (revitalization) amendments to the Constitution as well as the Anti-Communist Law of 1972,[7] actions that in turn led to the reorganization of social protest. The reforms provided for the election of the President by an electoral college along with the right for unlimited re-elections for six-year terms. The President also gained the right to name one-third of the national legislators. The essence of these reforms was meant to further concentrate power in the hands of the executive and to perpetuate its rule. Park, like Rhee, intended to remain in power indefinitely and the *Yushin* reform now allowed the presentation of a

---

7 Subsequently known as the National Security Law, it prohibits contact with families or citizens of North Korea as well as being in possession of any text, audio, or video material that favour or supports reunification. Under this law, about 80,000 items have been eliminated from the Internet and 82 people have been tried in court.

single presidential candidate in elections. This made the "elections" of 1973 and 1978 an even simpler matter and in order to support these authoritarian reforms, Park created the South Korean Central Intelligence Agency (KCIA)[8] that enabled him to extend his reign for 18 years.

By 1973, the newly re-elected Park was the target of protests by students and workers fighting for the construction of democracy. Park issued an official decree in 1975 that now considered any criticism of the political system to be a crime (Aguirre 2009: 30). Throughout the 1970's, the regime's violations of human rights intensified, civil liberties remained restricted and the political opposition was persecuted. Nevertheless, large protests continued to grow, leading to the restoration of martial law and the use of the military to quell the renewed opposition.

As defence mechanisms of civil society, Catholic Church leaders turned their spaces into zones of protection for the opposition and a spirit of *Minjung*[9] provided an ideological basis for the peaceful demonstrations being organized against the *Yushin* system. Nonetheless, worker protests, strikes and the formation of unions were harshly repressed by the CIAK throughout the 1970s as any union or any organized protest was seen as a threat to the national industrialization plan. This makes it clear that the aims of the bureaucratic-authoritarian state had little interest in improving the conditions of everyday life for the majority. In this climate of the absence of freedoms and rights for workers, women workers also became involved in actions to express their dissatisfaction, even as they were harshly repressed and many were fired from their jobs on the accusation of causing damage to private property. In one famous case, a textile worker, Chon Tae-il set himself on fire in a suicide protest against the poor working conditions that women and girls were being subjected to in the textile sweatshops of Seoul (León 2003: 55). Chon Tae-il's self-immolation inspired the formation of the Ch'onggye Independent Union of Garment Workers, most of whose members were women, as they demanded respect and recognition of human rights under the harsh conditions of repression. In the Dongil Textile Company, another autonomous union was formed but condi-

---

8   The KCIA is considered to be an alternative organization to the military, responsible for coordinating the tasks of intelligence and is known to have carried out persecution, torture and assassinations of oppositional figures to the Park Dictatorship, up until 26 October 1979 when KCIA Director Kim Jae Kyu assassinated President Park. See Díaz (2005).

9   *Minjung* roughly translates into "people's spirit" as it implies the construction of alliances among those marginalized from the benefits of economic development. The spirit of *Minjung* helped embolden the growing movement in opposition to state authoritarianism. See Koo (1993).

tions made it impossible to remedy the situation of women textile workers. The whole model of economic growth continued to chart upwards on the basis of low wages and poor working conditions.

Sonia Winer (2005: 309) argues that the repression and social discipline were possible thanks to the fluid communications that the KCIA had from the ground up, operating horizontally with the Federal Police and the Army Security Command. KCIA agents were scattered all across opposition political groups, newspaper offices, radio and TV stations, trade unions and universities. Despite all of the repressive measures imposed by the regime and the espionage of the KCIA, workers nevertheless continued organizing and went on to create the Urban Industrial Mission (UIM) that struggled to make workers aware of their rights.

In the economic sphere, Park continued to support the development and consolidation of the *Chaebol*. In order to strengthen them, he relied upon the nationalized banking system, low interest loans, selective trade policy products with preferential loans, and government participation in venture capital with private entrepreneurs, maintaining selective restrictions on foreign direct investment. In addition, the model of export-oriented industrialization (EOI) was planned and implemented.

According to Eric Toussaint (2006), the IMF questioned at one point whether this industrialization was too ambitious and issued a recommendation to curtail it. The South Korean authorities did not follow the IMF recommendations, however, as illustrated by the spectacular development program of heavy industry that was carried out between 1977–1979. For two years, the state dedicated 80% of all investments to the effort. Funding for the colossal effort was secured by a massive growth in debt on the part of the state and its bank, as well as of private companies, while the state turned to freezing all pension funds and the forced use of private savings (Toussaint 2006). It amounted to clear economic coercion on the back of the most defenceless classes of the population.

Throughout 1979, social movements that supported the overthrow of the regime and the installation of a real political democracy continued to mobilize, with university students in the vanguard. Their actions were harshly repressed by the authorities (León 2003: 57). Paradoxically, it was in that year that President Park was himself assassinated by the director of the KCIA, an institution that was created to protect his regime. He was succeeded by his Prime Minister Choi Kyu Hah who initiated reforms in an attempt to quell the protests, including the freeing of political prisoners, re-admission of dismissed students to the universities, and repealing constitutional restrictions that had prevented the organization of political parties and criticism of the *Yushin* Constitution. These tentative moves towards democratization notwithstanding, the bureaucratic-authoritarian state remained intact following the assassination of Park.

Choi's tenure as President was short-lived after a Coup was carried out in December, 1979 under the direction of General Chun Doo Hwan who assumed leadership of the KCIA and moved to dissolve the National Assembly, declare martial law, ban political parties, close the universities, and reinstate the prohibition of political activities while ordering the detention of student and union leaders all across the nation (Winer 2005: 293–294). From October 1979 to May 1980, workers, students and the middle-classes spearheaded a growing movement that demanded the end of martial law, respect for human rights, and the return to democratization. The ferocious wave of repression mounted to suppress it culminated in the Rebellion of Gwangju.

It was in Gwangju where elite troops were parachuted in, even possibly drugged, and dispersed throughout the city on a wave of indiscriminate killing of students, women, children and anyone who dared to show their face. As one analyst details the US support for the operation: "the forces of repression waited three days to enter Gwangju so that the US aircraft carrier Midway and other US naval vessels were able to reach South Korean waters" (Cumings 1997). The entire operation and deployment of army troops demonstrated the intervention of the United States in the background as Washington was helping to repress social movement resistance to capital. Even with armed persons among the resistance, there were no reported political revenge killings or attacks on banks or public buildings, looting or attacks on private property (Han 1998: 22).

Many student activists were killed in the slaughter of Gwangju, also known as "the Seoul Spring." But in spite of the regime violence, the marches and protest movements in the aftermath of Gwangju did not stop (Ogle 1991). The government continued to use the army to wage its violent oppression, students continued creating links with the working class, and in the absence of political parties the middle classes became bearers of social discontent against government authoritarianism. While the Gwangju massacre was officially declared to be an action against citizens under the influence of the Communists, it prohibited any public discussion of the tragedy (Díaz 2005). The fact that the mass struggle continued under the heightened repression helped to mark it as the beginning of the second stage of political liberalization (Penoncello 2005: 257).

President Chun, as head of a special committee to take measures related to national security, announced in April 1981 that new changes had been made to the Constitution. A presidential mandate would now consist of only one, seven-year term although some *Yushin* mechanisms would be maintained. This yielded an administrative decree for the dissolution of unions, particularly the

Ch'onggye Textile Workers, and the creation of strikebreakers known as the "white skulls" (Días 2005). In addition, he established new punishments against leaders, government officials, students, journalists, teachers, trade unionists and public employees who had participated in political activities, ordering them sent to "purification centres" in remote mountainous areas. From the very beginning of his mandate, those arrested on charges under the Law of National Security made up a third of all political prisoners (Cumings 1997).

Chun ultimately made himself President of the Fifth Republic with support from General Roh Tae-Woo who had also collaborated in the repression of Gwangju. Despite all of this, some of the international legitimacy that South Korea had in its purported democratic advance was because of its support of education as well as its emphasis on scientific and technological research. The implementation of the national plan on education had taken place in 1970. In 1980 the law created in 1972 governing institutions of higher education was modified to bolster research in universities with private investment and they subsequently were given greater financial support. For example, the National University of Seoul received an increase of nearly 80%. In 1985, the Presidential Commission for Education Reform was created with the objective of analysing the scope and limits of the quality and competitiveness of schools (Rubio 2013: 28–29).

Throughout these decades of military authoritarianism, South Korea emerged as an industrial power with the formation of big capital. Enrique Valencia (2001) argues that in addition to its role in the anti-communist struggle and national defence, the state hid behind a veil of nationalist legitimacy. It was portrayed as being at the helm of a project for national industrialization objectives, with gradual improvement in the real wages of workers, a growing stability associated with "a job for life," and greater levels of household consumption.[10] Excluded from this package were electoral legitimacy and the coercion of citizens to accept an implied waiver of their civil, social and political rights, their equality before the law, social discrimination and their freedom of thought.

As protests continued under the Chun presidency, there were particularly large uprisings in the cities of Pusan and Masan that were crushed by the army. The strategy of the mass movement rested upon workers, students and the church establishing an alliance of mutual support in pressuring the regime. In addition to political opposition, they constituted a core of resistance that

---

10   The practice in the *Chaebols* of employment for life came to end after the crisis of 1997 and the turn towards more flexible employment practices. See Valencia (2001: 103).

helped boost the mass mobilizations into June of 1987. In one protest by students at the University of Yonsei, a protester, Lee Han Yeol, was seriously wounded when a tear gas grenade exploded near his head. He later died on July 5. More than 1.6 million people attended his funeral held on July 9 that was to become a symbol of national protests. The government was left with only two options: either to continue its violent repression using the army or to initiate a real opening towards democratization.

With respect to economic growth, the year 1980 marked for the first time a slight downturn in the GDP to 5.2% and an increase in the annual inflation rate to 3.8%. This situation began to make it clear that the protectionist state strategy was no longer the most efficient way to favour capital, leading to the onset of reforms for economic liberalization. This would include the privatization of banks, a hike in interest rates, greater reliance on foreign loans, and opening up the economy to direct foreign investment in industry and commerce. Not long after these adjustment policies took hold, the working class was bearing the brunt of the costs as wages stagnated while layoffs and unemployment increased.

Genevieve Marchini (2009) offers an overview of how the sequence of economic reforms unfolded over the decade. First, the liberalization of trade and institutional diversification of the domestic financial system under the tight control of capital was accompanied with the privatization of the banking system, even as it remained under control of the authorities. These would provide companies with loans under rates set by the state. Second, South Korea had resorted to international borrowing to finance its growth in the post-war era, beginning with a prudent process of financial liberalization, initially limited to a partial liberalization of the outflows of South Korean capital in the context of current account surpluses. However, the international financial reforms of 1990 coincided with the liberalization of the domestic banking system and the opening of the portfolio of investments, actions that would make the South Korean economy more vulnerable to any external fluctuations, especially if negative.

Socially and politically in 1985, the mass movement of students and workers resumed its struggle, including the mobilization of a successful strike at the Daewoo Motors and General Motors conglomerate where an independent union was formed. Meanwhile, the movement made demands for constitutional reforms to eliminate the Electoral College imposed by President Park and the restoration of presidential elections with direct suffrage. In short, they were demanding a democratic opening and respect for civil rights. In May 1986, these demands intensified when 166,000 students participated in a mass demonstration. University students had by then come to constitute the majority of "political prisoners," about 800 among 1,300 detainees (Toussaint 2006).

Towards a New Constitutional Framework

To appease widespread social protests, President Chun called elections having just recently formed the New Democratic Party of Korea and appointed Roh Tae Woo his successor. The elections could freely elect representatives to the National Assembly and Chun promised further democratic reforms having acceded to the main demand of direct presidential elections and recognition of the political, civil and economic freedoms (Valencia 2001: 105). In the area of social rights, students who had been expelled could now return to their universities. Chun soon after won a diplomatic victory when the International Olympic Committee designated Seoul as the venue for the 1988 Summer Olympic Games.

On 10 June 1987 at the convention of the Democratic Justice Party in the Jamsil Stadium, the military government suddenly declared that Roh Tae Woo was its presidential candidate. The nomination was seen by the public as a process that was deferred and belated, a final affront to their demands for reforms to the Constitution to guarantee direct presidential elections. His appointment ignited a lot of public animosity, resulting in large social movement mobilizations. First there were student demonstrations that were soon to be joined by the Popular Union for Democracy and Unification and the Mutual of Workers. People from all walks of life took to the streets in 22 cities to demand the holding of direct presidential elections. On June 18, a national demonstration was held to demand in addition to constitutional amendments that the use of tear gas grenades be prohibited. This event brought 1.5 million people to the streets in 16 cities and for the first time even gained support from some bureaucrats (Díaz 2005). In response, President Chun ordered on June 19 that the army be mobilized against the protests.

When President Chun issued these orders to mobilize the army against citizens, he hesitated to deploy them fearing a repetition of what happened in Gwangju. Millions of workers proceeded to go on strike to demand improvements in working conditions, pay and social security (Valencia 2001: 104). On June 26, the Grand National March for Peace was convened "Guk-bon", with over one million participants from 34 cities, leading to the arrest of 3,647 protesters. The popular movement for democracy ended with huge demonstrations and strikes throughout July 1987. This movement, beginning with its inception in 1985 until its completion in 1987 marked the third stage of transition to democracy (Penoncello 2005: 258).

On June 29, Rho responded to the movement's demands and declared that he would carry out a broad program of constitutional reforms. He specifically agreed to respect the right for presidential elections by direct popular vote as well as to free political opposition leader Kim Young Sam,[11] who along with Kim Dae Jung, assumed leadership of the newly created opposition under the auspices of the New Democratic Party of the New Korea. The opposition leaders were, however, unable to draw up a consensus list of candidates and arrived in the elections with a divided vote among competing candidates. After a relatively clean election campaign, Roh Tae Woo won the election. The victory made him the first democratically elected President and this symbolized to the larger society that the transition to direct presidential elections was something of a moral victory for the opposition. The government proceeded to institute further reforms that led to the establishment of the Sixth Republic.

Roh Tae Woo (1988–1993) exercised his mandate by making structural changes to the economic model. He promoted the export of technology-intensive products and implemented a commercial conversion policy that moved along two paths: first, the volume of exports was increased with planned annual average expansion rates of 5.7 % (to 2010); and second, growth in imports was restricted from the annual average of 6 % to 2 %, this in order to move the economy towards achieving a trade surplus.

Meanwhile, political demands continued to be mounted against the government. A total of 297 groups formed the National Emergency Committee for the Enactment of a Special Law to Punish the Ideologues of the May 18 Massacre. Their campaign managed to collect more than a million signatures of support. The investigation into the event at Gwangju in 1980 confirmed the responsibility of former President Chun. However, President Roh announced that he would not file charges against previous presidents so as to not damage "national unity." The social movement response to the impunity was swift as the "Korean Council of Professors for Democratization" launched a national campaign to collect signatures to demand a law to prosecute military leaders (Ho 1998). Similarly, the June Democracy Movement spread throughout the country, calling for mass protests that occurred from 10 June through 29 June 1988. Given the intensity of social demands being made, the repression by the army was no longer practical and the national situation was ripe for democratization.

---

11   Kim Young Sam, leader of the New Democratic Party, first arrived in the National Assembly in 1954, and was expelled from it in 1979 and placed under house arrest for two years. In 1980, he was prohibited from participating in any manner with any form of social protest or political participation for having criticized the government for being an authentic dictatorship that failed to respect civil, political or human rights. In 1987, his rights were restored and he ran for election.

During the First through the Sixth Republics, the state's bureaucratic authoritarian political regime had directed the process of economic development under a nationalist, industrialist and anti-communist ideology. Up until the Roh administration, the market and pricing mechanisms were used to regulate and change the behavioural patterns of capital and labour, always in favour of capital. The priority of industrialization was largely successful in generating economic growth, even as much of society remained impoverished and immersed in conflict and political instability (Valencia 2001). One of the principal factors contributing to this conflict and instability remained the exclusion from political participation and the failure to fully recognize the rights of citizenship.

The economic development strategy that was initiated and consolidated during the dictatorship of Park Chung-Hee and which remained in place up through the government of Chun Doo-Hwan can be characterized as a mobilized economy. In this economic model, workers were expected to make sacrifices and comply through their obedience at work in exchange for occasional modest social concessions when determined apt from above. In analysing the political aspects of these governments, it can be observed that while the authoritarian presidents created a number of institutions and figures representative of democracy, never in their 30 years in power were these mechanisms put into practice.

Transition to Democracy, Civil Governments and Economic Crises

After many years of protests, movements and struggle for democracy, it can be said that it was President Kim Young Sam (1993–1998) who presided over the transition as the first civilian directly elected by popular vote. Kim's interest in politics dated back to 1954 when he won a seat in the National Assembly on account of his call for economic and political reforms. As a result, Kim was expelled from the Assembly and forced into exile abroad. Kim returned to the political scene as part of the opposition in 1987. A huge surprise for all his followers was that in 1990, while arguing in favour of equal political rights and reforms that could put an end to the conflicts between parties, he allied with President Roh and Kim Jong Pil, the last president of the New Democratic Republican Party. This alliance resulted in the creation of the Democratic Liberal Party that with the support of workers secured the future presidential victory of Kim.

In the legislative elections of 1992, the Democratic Liberal Party scored a very low turnout and the need for strong leadership led the party chairmanship to be turned over to Kim Young Sam. Kim was named the party's candidate for the presidency with the slogan "Building a New Korea" (Romero 2001: 8). His campaign attracted voters on account of his proposed economic reforms that would remove the military from politics and attack corruption within the state bureaucracy, the armed forces, and in the relationship between the government and the *Chaebols*. In June 1995, the first free local elections were held although the results for the Democratic Liberal Party were poor, losing 10 of the 15 metropolitan districts, thus showing that the unification of parties was still not fully matured. In the meantime, revelations of corruption that reached up to former presidents Chun and Rho did further damage. This became a determinant factor in the November 1995 demonstrations in most major cities across the country, which demanded the prosecution of those responsible for the massacre at Gwangju.

Faced with intense popular pressure, the Kim government moved in late 1995 and early 1996 to put the former presidents on trial on the charges of corruption and diversion of resources for which they were convicted.[12] But the success of the movement was diminished when the government also moved to pass a package of reforms that included weakening the unions and taking away from workers some of their recently obtained rights, while at the same time granting more power to the KCIA. These moves affected the voting as evidenced by the disaffected workers, students and women who opted for opposition candidates (Bavoleo and Ledevito 2009).

In the economic sphere, the 1993–1996 shift in the conduct of economic policies quickly fell in line with the requirements of neoliberalism as portfolio investments and foreign borrowing policies were privileged in order to obtain foreign capital (Marchini 2009: 99). In order to offer domestic credit to businesses, commercial banks and investment banks quadrupled their foreign liabilities, and maturities operated by banks to recover debts were transformed from short-term to medium-term periods. Greater domestic exposure of the *Chaebols* combined with poor regulation and supervision to facilitate the acquisition of international loans in foreign hard currency, thereby accelerating the vulnerability of the economy to shocks in the international economy. This situation would also weaken the moral credibility of the presidents who formed the opposition that had fought these policies.

---

12   The trial against the former presidents help to strengthen the Party of the New Era (which arose from the defunct Democratic Liberal Party) in the legislative elections of 1996. It gained an absolute majority of 220 of the 299 seats in that year. This notwithstanding, the privileged position of the President was affected by the arrest of his son, Kim Kyun Chul, who was accused of bribery and tax evasion. See Romero (2001: 8).

It was in this context of economic development that 1997 witnessed the eruption of the Asian financial crisis. South Korea was hit particularly hard since from 1993 it had resorted to orthodox neoliberal policies to sustain economic growth. These included: domestic market liberalization; deregulation and reduction of the state apparatus through privatization; deepening trade liberalization and direct foreign investment; and adopting trade legislation in accordance with the stipulations of the World Trade Organization, all of which exerted tremendous pressure toward achieving development goals and maintaining political control.

Along with trade liberalization, the regulations over foreign direct investment had also become liberalized which meant that the state was losing its direct control over the economy that it formerly had through the *Chaebols*. Indeed, the first action taken was to authorize the privatization of the *Chaebols*. In the area of monetary policy, the fixed exchange rate was abandoned in favour of flexible rates, while laws were enacted and tax exemptions granted in order to promote the creation of new industrial groups and attract greater direct foreign investment. In addition, South Korea's admission into the World Trade Organization (WTO) in 1995 marked a broad opening of trade with the outside world. In 1996, South Korea was recognized as a newly developed country and was accepted into the Organization for Economic Development and Cooperation (OECD), further intensifying the economic liberalization of the national economy.

At the same time, the drive to sustain economic growth under neoliberal orthodoxy led to ending restrictions on the inflows of direct foreign investment, thereby opening a channel for uncontrolled foreign exchange earnings that increased from US$788.5 billion to nearly US$1.8 trillion dollars in 1990, an increase of 125.2% with a corresponding appreciation of the national currency. This situation also brought about an imbalance in the trade balance and current account, such that the nation's trade deficit exceeded $10 billion in 1995 and $20 billion in 1996 (Gomez 2003), creating a ticking time bomb that went off the following year in 1997. It was then that the fall of the Thai Baht triggered the flight of foreign capital from its portfolio investments and bonds in a context where the currency exchange markets had already been liberalized.

Under these circumstances, the financial opening to capital and trade had increased investment expectations, but also stimulated domestic consumption, which in turn had expanded the trade deficit in the balance of payments where weakened government controls over current assets had substantially increased the vulnerability of the country to external financial shocks. At the same time, the *Chaebols*, no longer under state protection, had resorted to extensive external borrowing in the face of the economic downturn, leading to widespread charges of corrupt contracting, imprudent lending, excessive expansion, low profitability, high indebtedness, and a lack of transparency. As a result, the eruption of the Asian Financial Crisis in 1997 left the country hard hit and smothered under a commitment to further deepen its neoliberal structural adjustment reforms in exchange for receiving a bailout loan of $55 billion from the International Monetary Fund.

In the wake of the 1997 crisis, South Korea's gross domestic product contracted in 1998 by 6.7% in real terms, and by 1999, the unemployment rate had risen to 8.6% while international reserves fell to $3.9 billion (Gomez 2003). To confront the crisis, the strategy adopted by the government employed a combination of macroeconomic variables that involved market, fiscal and monetary policy mechanisms. Restrictions on imports were now eliminated. Import and export controls were abandoned in favour of tariffs as the main instrument of trade policy, accounting for 6.5% of tax revenues. A managed exchange rate allowed for economic stabilization without inflation while privatization was deepened and productivity increases were promoted. A central cause of the economic crisis of 1997–1998 was that deregulation of the financial sector took place without proper monitoring standards, creating a high level of vulnerability to domestic and external shocks. Meanwhile, although transformed as part of the 1997 IMF bailout, the *Chaebol* remained gigantic, forming almost parallel states that retained close connections to political power.

One factor not anticipated in the move towards democratization was unemployment, which since the 1997 crisis became a highly visible feature evidenced by the reduction in the hiring of young graduates. College graduates swarmed to job fairs looking for work even while their already completed exams were still pending. The government set up funds for these students to receive further training while waiting for companies to offer them job opportunities. Research projects were also launched in order to temporarily hire the students. In 1998, these projects were expected to create jobs for 450,000 graduates who nevertheless were forced to fiercely compete for them given the high demand and low supply of openings, and the fact that the targeted number of positions was not actually achieved. According to estimates, only 28% of the 195,000 graduates in 1997–1998 were placed, leaving out 139,000 who became added to the 230,000 already existing idled young professionals. This led to a proliferation of uncertainty on the part of young people and their families who continued to look to education as their ticket to a better standard of living (Romero 2001: 18–19).

In the field of science and technology, researchers made Herculean efforts in the areas of electronics, semiconductors and biotechnology to generate and refine technological advances. In 1997, the Review Committee on Science and Technology was replaced by a Ministerial Council on Science and Technology. The Council was chaired by the Deputy Minister of Finance and Economy and coordinated by the MOST. Beginning in 1998, the head of MOST assumed the directorship of the Ministerial Committee, the purpose of which was to more effectively manage the national scientific system. In addition, the government began to establish alliances as part of its efforts to become a country leader in the development of technology.

In this situation of crisis and the indifference of the political parties, the activities of the social movements focused on achieving and consolidating social reforms that represented public policy alternatives. Their actions revitalized the electoral climate. An example could be seen in the activities of the "Council of Civil Movement for Fair Elections" representing 50 civil society groups dedicated to monitoring the electoral process and the promotion of political participation. This coalition demanded and ultimately collaborated on the revision of electoral laws and the development of reports that established comparisons between the campaign rhetoric of candidates and the actions that were actually taken (Bavoleo and Ladevito 2009).

In December 1997, presidential elections were held. Kim Dae Jung (1998–2003) was the centre-left candidate proposed by the National Congress Party for New Politics, a party that had been recently formed in 1995. He was elected President with 40.3% of votes against 38% of his closest rival. His political campaign rested on two major proposals: Economic Structural Reform (being demanded by the IMF) and National Reunification. Politically speaking, his victory was the first real democratic change in the country and this marked the fourth and final stage of the struggles and movement for full recognition of political citizenship (Penoncello 2005: 261).

One of the proposals of his campaign consisted of renovating and restoring the solvency of the economy. Kim urged citizens to support national unity and get behind the reform since the plan aimed to salvage the IMF rescue credit, thus continuing an economic policy firmly circumscribed by the guidelines and austerity being imposed by this multilateral agency. Following this plan would mean liquidation of insolvent financial companies, putting a decisive end to protectionist practices, raising taxes and interest rates, accepting constraints on growth, further currency devaluation, further reductions of consumption, and the additional loss of jobs in a context where mass unemployment had already reached 2.1 million.

Acting on the other major thrust of his campaign, Kim Dae Jung went to Pyongyang, North Korea on 15 June 2000 to meet with his counterpart, President Kim Yong Il where agreements were made in principle to establish diplomatic relations. During that year, two milestones marked the initiation of the reconciliation process. First, in the opening ceremony of the Olympic Games in Sydney, Australia, both delegations marched under one flag. Second, Kim Dae Jung received the Nobel Peace Prize, even though the spirit of national reunification soon became diluted when threats of war came from North Korea.

In the electoral sphere, the "Citizens Alliance for the General Elections of 2000" comprising 412 civil organizations launched a campaign against those candidates deemed "unfit" for public office on account of anti-democratic acts, corruption, tax evasion or other illegal or immoral activities. In retrospect, 86 of them got eliminated (Romero 2009).

Meanwhile on the economic front, the IMF had predicted that if its prescribed orthodoxy were properly applied in South Korea, i.e., fiscal austerity and monetary tightening (GDP had fallen to 6.9%), that would allow it to eke out a 2% growth for 1999. Its orthodox prescriptions were nevertheless discarded in favour of more heterodox strategies, including a fiscal deficit set at 3% of GDP in 1999 and financial liberalization including a more flexible approach to direct foreign investment. These policies achieved a growth of 9.5% for the year and 8.5% for 2000 as exports and imports increased, respectively, at 8.6% and 28.5%. This enabled a current account surplus of 5.5% of GDP in 1999 and 2.4% of GDP in 2000, allowing for continued growth in subsequent years (Guimaraes 2010). In short, the South Korean economy emerged victorious.

South Korea's economic policy had focused on turning the nation into a highly industrialized economy, making education a priority. Despite the crisis, the budget for education was increased, although higher education operates in a scheme shared by the state and the families of students. The rate of university enrolment went from 10% in 1980 to 70% in 2000 while the investment ration in research and development went from 0.7% in 1980 to 2% in 2000 (Guimaraes 2010). These figures indicate that education and research policies were considered by the government to be a binomial for building creative capacities, in particular, since they allowed the nation to transcend the status of being a technological imitator in favour of being an innovator for its products.

A new law was unveiled in 2001 to help manage more precisely the national scientific and research effort by implementing a legal and institutional basis for a coordinated policy in support of science and technology. The law's purpose was to introduce a new social vision and promote the formation of resources and regional development. Despite the economic crisis of 2008 that shook the country, it invested about 3.1% of GDP that year, helping it to reach a 4.18% share of all patents registered in the world as it registered 9.7 researchers per 1,000 of its economically active population.

Other factors, however, included a fall in real wages, increased job inse-
curity, a growing crisis in its model of labour relations, and deepened social
inequalities that increased the incidence of poverty (Valencia 2001: 110). The
new democratic government could not disentangle itself from the crisis and
was perceived throughout the population as responsible. On the other hand,
the President adopted neoliberal language in his discourse and abandoned his
earlier criticism of the policies imposed by the IMF that further narrowed the
social legitimacy of the democratic transition.

The Institutionalization of Democracy

Roh Moo Hyun (2003–2008) was nominated presidential candidate by the Millennium Democratic Party with the image of an honest politician with working
class origins and sensitivity to the everyday concerns of citizens. He was free of
regionalist party patronage and the stigma of clientelism and won with 48.9%
of the popular vote against 46.6% of the opposition. Once installed in power,
Roh moved towards administrative decentralization and at the same time pursued close trade ties with the United States. His goal was to turn the country
into a logistics business emporium in Northeast Asia. A significant democratic
feature of his approach was his resumption of a policy of rapprochement with
North Korea. In 2007, he signed the Declaration of Peace and Prosperity in the
Inter-Korean Summit, a move somewhat diluted by the political game of economic interests that he was playing beyond the Korean borders.

In terms of economics, he implemented liberal reforms to regulate and
reduce the power of the *Chaebols* that traditionally were immersed in bribery
and corruption. This web of accumulated corruption was identified by the IMF
as a main obstacle to implementing a true market system of free competition,
undistorted by the collusion of political and economic private interests. In this
context, the President sought to end a long-standing practice in the *Chaebols*
that involved the transmission of corporate ownership from parents to children
or among relatives as if they constituted an estate inheritance. So he under-
took measures requiring a periodic replacement of *Chaebol* directors. Roh also
focused economic policies towards reforming the tax system and establishing
administrative transparency through the adoption of international account-
ing standards. In addition, he promoted greater openness to direct foreign
investment by establishing financial regulations to increase the participation
of minority shareholders.

A situation that alarmed authorities arose between 2006 and 2007. The
Chaebols had contracted huge debts to the point where the foreign debt had in

just two years grown from 23.7% to close to 40% of GDP, almost half of which was based on short-term repayment. The character of the debts, nevertheless, differed from those contracted before 1997, so that up until mid-2008 many financial analysts emphasized the low level of risk posed to the national economy. A relatively high proportion of these debts, (i.e., a little more than a third) were backed up in a term of three to four years by export earnings. In fact, the debts were contracted by banks that acted as a counterpart for exporters in derivative transactions (hard currency futures) by which companies sought to protect their future income from the appreciation of the Won (Toussaint 2006). On the other hand, a portion of the debt consisted of loans from banks operating in South Korea and the fraction of the debt contracted by local banks, considered riskier, was smaller and supported entirely by short-term assets in foreign currencies.

Due to the way in which the economy was operating, President Roh and his party began to lose popularity. His representatives in the National Assembly soon abandoned the Party and so in 2004, the President founded "Our Open Party" and just managed to maintain the weak presence of his party loyalists in the Assembly. Rival parties in the National Assembly harnessed enough votes (but without a popular plebiscite) to unseat the President from office. The manoeuvre was greeted, however, by more than 70 thousand people who took to the streets of Seoul to protest, while the Constitutional Court approved the measure. For six months, the street demonstrations and night-time marches with lighted candles continued as intellectuals, trade unionists, students and NGO members joined. Citizen anger against the political class was expressed in a statement made by civic groups to the effect of: "Congressmen cannot throw a President out of office who was elected by the people without the consent of the people just because they may hold a majority in the National Assembly. This is an outrage that was committed by the political powers of the past" (Ilbo 2004).

In April 2004, elections to the National Assembly were held in which a reconfiguration of the parties in power occurred. The Grand National Party who had promoted the impeachment of the President lost their majority while the Millennium Democratic Party lost most, going from 61 to 9 seats, where 11 deputies were needed to form a parliamentary group. In contrast, the Democratic Labour Party rooted in the trade unions reached 10 seats and Our Open Party obtained a majority 152 of the 299 Assembly seats. For the first time in history, the political left wielded a majority in the National Assembly (Leon 2006: 65–66). This showed once again how the economic crisis and reigning style of managing the economy impacted the political mood of the public.

Lee Myung-bak (2008–2012) won nomination of the Grand National Party following a tight internal competition against Party Chairwoman Park Geun-hye. During his campaign, he convinced both the employers and unions that had been punished by the structural changes to vote for him with the promise to restore economic growth back to 7% and to create jobs (Giné 2010). With Lee's subsequent electoral victory, the conservative Grand National Party regained some of their previous power by obtaining a majority in the National Assembly in the elections of 9 April 2008 with 153 of 299 seats.

As the former president of Hyundai Construction, where he worked for over 27 years, Lee believed that the unions had mishandled their freedom by calling strikes that paralyzed economic activity. During his election campaign, Lee used the slogan "smart market economy" to convey the idea that competitiveness, freedom and creativity must be protected. He argued for a strong "empirical pragmatism," maintaining that any policy guided by ideology should be avoided and that the national interest should always remain the top priority. This notwithstanding, the low voter turnout (46%) indicated considerable scepticism on the part of young people regarding the electoral promises of the political establishment.

Lee Myung-bak wanted to show that South Korea had reached maturity not only in an economic sense but in the political realm as well. In his first official visit to the United States on 18 April 2008, he opened the South Korean market to American beef even though this decision was not ratified by the United Democratic Party in the National Assembly or approved by public opinion. In 2007, however, President Rho had established commercial trade agreements with the United States that included a Free Trade Agreement. Then Senators Barack Obama and Hillary Clinton both expressed their intention to renegotiate some aspects of the Treaty that harmed the interests of US agricultural and automotive sectors. In the National Assembly, the festering parliamentary conflict continued until March when the Grand National Party and the United Democratic Party reached an agreement to clear the gridlock and approve the economic legislative proposals that would allow the Free Trade Agreement to go into effect.

With the collapse of the Lehman Brothers investment bank in 2008, the South Korean economy, now dependent on direct foreign investment like Japan and China, bore the impact of the US and European Union financial crisis. Industrial production dropped precipitously and investment declined by 14% while the damage to the labour market resulted in the loss of over 100,000 jobs, creating a snowballing effect in the reduction of domestic consumption. This now forced the South Korean government to return to those pending structural reforms still unfulfilled as recommended by a report of the OECD.

The crisis of 2008 further showed that the revitalization of the domestic market had its limits. The economic strategy had triggered domestic consumption and the expansion of the domestic market through the issuance of credit cards. Prior to 1999, about 16% of consumption had been supported by credit card purchases and by 2002 the percentage had reached 56%. As this policy was doomed to fail, the consumption rate in 2003 reached –1% while average household debt reached US$25,000, leading many credit card users to declare a moratorium on payment and about three to four million fell into bankruptcy (León 2005). In general, the economy entered into a precarious state as it became ever clearer that the institutions created to provide social security were mostly symbolic, rather than solvent, and attention to the population was selective. Both the social security and education systems rested upon a system of shared costs between the state and family, making it highly expensive for the latter.

The electoral victory of Park Geun-hye (2012–2016) marked an important step in the field of domestic policy, relations between the two Koreas, and Asian security and cooperation. Park Geun-hye cantered her campaign on two demands: national unity and democratization of the economy. On the economic side, she proposed a program called "creative economy" designed to diversify and consolidate the entrepreneurial progress and social welfare. This translated into limiting the power of the *Chaebols* and supporting the development of small and medium sized firms while extending the scope of social services, increasing employment opportunities for young people, ensuring the rights of children, and fighting the entrenched corruption of the political class. She sought to reinforce this economic program by extending foreign commerce through the establishment of new free trade agreements with major countries and economic areas of the world.

In 1997, Park was a member of the New Korea Party that until 1995 had been called the Liberal Democratic Party. This in turn had resulted from the merger in 1990 of Roh's Party for Justice and Democracy, the Party of Democracy and Reunification of Kim Young Sam, and the New Democratic Republican Party of Kim Jong-pil. The Party of Justice and Democracy had taken the structure of the Party of Democracy and Reunification founded by Park in 1960 and the New Korea Party founded by Chun in 1980. So by carrying out this merger, Park had reconnected with her political past. In April 1998, the votes she obtained in Dalseong, Daegu County, gave Park the opportunity to become a legislator in the National Assembly.

In 2000, the electorate had renewed her mandate (61.4% of the vote), beating the former President Kim Dae Jung. In the same year at the University of Daegu, she indicated that she would like to be the candidate of the Grand

National Party in the 2000 elections, but upon entering into a conflict with the party chairman, she abandoned that plan and formed the Korean Coalition for the Future, later creating the New Frontier Party, taking it to power with an electoral victory with a 75.8% level of participation.

Park's government had many resources to boost growth and promote the country as an attractive site for investment. South Korea's workers are among the most competitive in the world for their educational training, dedication and discipline in a country that has adopted English as a second language. Companies like Samsung, Hyundai and LG manufacture products sought all around the world, like plasma and LCD Screens, mobile phones and many other appliances. It is in first place in the manufacture of ships, third in the production of semiconductors, fourth in digital electronic products, fifth in the development of the textile, steel and petrochemical industry, and sixth in automobile manufacturing.

South Korea possesses a strong, high-tech infrastructure that is one of the most advanced in Internet and communications development. It has a high penetration of mobile broadband (91% overall) and fixed mobile broadband (36.6%). The Internet connection in the country is among the highest world-wide with a 1Gbps rate in 2012. Since 2003, robotics formed part of the major research and development projects. In 2005, the Institute of Science and Advanced Technology developed the second humanoid robot in the world: HUBO. In 2006, the Institute of Industrial Technology developed the first Android: Ever-1, and have since created even more complex models. Up until 2012, the National Technology program was a priority and it now forms part of the Twenty-First Century Frontiers Program.

Conclusion

Since the founding of South Korea, the nation has been referred to by some as an example of democracy. As described in this work, however, the repressive and authoritarian policies imposed by the bureaucratic-authoritarian state over a 40 year period does not represent what should be an authentic model for democracy. In recent years, it has initiated changes that point to a transition to democracy that is still being consolidated.

Under the construction of a model of export-oriented industrialization, South Korea has endured a permanent policy of repression of social move-ments (e.g., ban on trade unions), including against students and legitimate student protest (e.g., expulsion of activists from universities), alongside a super-exploitation of peasants and workers. The South Korean state was erected as the driver for economic development, always while under the benevolent watch of the United States that tolerated there what it declared to be intolerable in other countries.

The initial industrialization of South Korea did not depend at all on foreign loans or foreign investments, so that instead of applying the dogmatic prescriptions of the IMF, the South Korean leadership opted for designing their own development strategy. Their alternative policies featured import substitution combined with an aggressive promotion of exports, all supported by strong state interventionism as a planner, driver and promoter of development.

South Korea exemplifies a presidentialist regime that has assimilated some formulas of parliamentary democracy. If a President does not have an absolute majority, he or she may appoint an intermediary responsible for the development of administration while they dedicate themselves to state policies. Meanwhile, the party system and parties have displayed little ideological differentiation and a low level of institutionalization. The parties have operated in practice as lax organizations, with little discipline and a regionalist character, maintaining an excessive presence and influence of economic and industrial groups in electoral matters which encourages corruption. To the extent that the regime of parties improves, there will be further progress in the strengthening of democracy and in consolidating full respect for the rights of citizenship.

Beginning in 1987, a second period opened in the capital-labour relation. One of the main functions performed by the authoritarian bureaucratic state is now history. It achieved the necessary accumulation of capital by suppressing workers, tightly controlling their wages, and repressing organized social movements' resistance. There is recognition of the relationship today, not as equal partners, but as a social relation where the working class has become aware that their work is creative and at the same time is amassing capital.

Up until 1997, the population had full confidence in the future. Beginning in 1998 with the breakdown of the economy came uncertainty, and although the economy was able to recover from the crisis and pay off its debts to the IMF, the government continued to make progress in recovering the confidence of the public. In more recent years, the economy has hovered around full employment with growth in 2012 at 3.5%, increasing to 4% in 2013. There has been a very low level of public debt and the nation retains reserves in excess of US$ 300 billion.

The government faces various challenges in the economic sphere. The first is the implementation of economic development strategies that favour the freedom of capital flows, an essential ingredient for making the country a financial hub for East-Asia. It must also manage to isolate private investors from the flow of easy money, thus avoiding that the development strategy be swayed by fluctuations and international monetary conditions (Giné 2010). Currently, the main instrument of the economy is the use of tariffs and although there is a great disparity in their application, it provides up to about 6.5% of tax revenues. Moreover, the opening to foreign direct investment in order to attract capital has had the objective of guiding it into companies where they can crystallize technology transfer and help build domestic technological capabilities.

Being a highly industrialized country with high population density, it has allocated resources for advancing moves towards a green economy. It is in the field of clean energy where local authorities are promoting development. One of the institutions most recently created was the Ministry of the Environment that has been responsible for public policies and a legal framework of protection in order to improve environmental management, mainly over the care of water and air. In this field, researchers are seeking answers that can further affect the implementation of alternative and renewable energies (Giné 2010).

In order to overcome the crisis, it was necessary to put aside the protectionist approach and open up to foreign markets. Although the country had free trade agreements, they were not quickly put into place. But since 2003, the move for additional free trade agreements resumed and several were signed with other countries. South Korea now has agreements with the European Union (1996, 2001), United States (2007), members of the Association of Southeast Asian Nations (2007, 2009), India (2009) and in Latin America with Argentina, Chile, Colombia, Peru and Mexico. South Korean companies can export goods free of tariff barriers to economies that together account for 60% of the global GDP, seeking to become a global platform for business and free trade.

The government has negotiated agreements with China and Japan with the idea that companies around the world are installed on its territory so that from there, there is easy access to major Asian markets. Aware that it cannot compete with China and other Asian countries on prices, it has opted for technological innovation, quality and knowledge as its survival strategy. It therefore invests in research and development at the level of 3.74% of GDP while it continues to strengthen its education system.

In the Twenty-First Century, educational policies are at the forefront of global changes. According to the Ministry of Education, the "Brain Korea 21 Project" has worked to strengthen research and competitiveness in universities with the vision that graduates be of high quality and able to provide an innovative workforce for industry in local regions. The Connect Korea project promotes links between universities and industry, and the establishment of consortia partnerships between the university and technology licensing offices. It also implements public policies for quality assurance of higher education (evaluation and accreditation, budgetary allocation based on external evaluation, etc.) and for providing financial incentives for reform and restructuring that will grant increased autonomy to public universities.

In political terms, the system of political parties corresponds to a moderately multi-party democratic system that displays a low level of institutionalization, ideology and political projects. Leaders continue to transcend their party and this has weakened some of them due to their frequent reorganization. The institutionalization of a true party system can function as a hub for consolidating democracy and countering corruption and patronage. However, the facts and practices have shown that not to be the main determining factor. The civil society, including students, teachers, workers, women, and others, have made demands analogous to political parties, though they cannot replace them, so there remains a pending task of their consolidation, institutionalization and improving their quality in order to develop a more solid way of building a more democratic political regime.

CHAPTER 7

Latin American Democracy as an Alternative Work in Progress

Ximena de la Barra and R.A. Dello Buono

There are many ways of understanding democracy in whose name the greatest outrages are often committed. For the ancient Greeks, it meant 'government of the people.' For Montesquieu, democracy implicated the *trias politica* or the division of powers of the state to establish equilibria over authority. However, for contemporary Latin America it has come to mean only the right of everyone to cast a vote freely. In reality, this right has never been fully exercised and worse yet, voters often have no real options. When the struggle for democracy takes place in a regional context of almost continuous foreign interventionism, the results, just as in antiquity, tend to evolve towards ever more sophisticated forms of domination.

From the overthrow of the democratically elected government of Prime Minister Mohammed Mosaddegh and his National Front cabinet in Iran in 1953 to the overthrow of the first democratically elected president in Guatemala in 1954, Jacobo Arbenz, the United States had almost exclusively relied on the Central Intelligence Agency (CIA) and its many 'front' agencies to orchestrate desired political changes outside its borders. A few years later during the aggression waged to combat the nascent Cuban revolution, this Agency took further shape and grew to become the ideal instrument for the practices of 'dirty war.' Since then, its clandestine, criminal actions have served to impose Washington's order on multiple occasions, performing the most unseemly aspects of US foreign policy, including psychological warfare and propaganda operations in addition to paramilitary operations. It's worth noting that the CIA was not only composed of paramilitaries, or agents licensed to kill. Clandestine operations were prepared by well-known political figures, some of whom are still active today such as John Negroponte (Calvo Ospina 2010).

Since the early 1980s, the resurgence of neoconservatism or neocon ideology in the United States championed a renewed commitment by Washington to extend its version of democracy throughout the Americas. This took place during the two consecutive administrations of Ronald Reagan (1981–1988) and continued unchanged in the administration of George H.W. Bush (1989–1992). A broad consensus was reached in the governing clique of the United States to implement a foreign policy through those political and ideological mechanisms that could guarantee hemispheric control through more appropriate 'democratic' institutions. The process began with the shadow of the Carter Doctrine still projecting on Washington, leaving many military dictatorships previously installed with US support in a position of 'official discredit.' The new model coming into vogue was virtually inspired by contemporary political theorists such as Robert Dahl. In the 1970s, Dahl introduced the concept of 'polyarchy' emphasizing the importance of competitive procedures for power among aspiring leaders. Dahl argued that these kinds of procedural mechanisms 'ensured' the compenetration of leaders with their constituencies (Dahl 1971). The kind of democratic elitism to which Dahl, among others, lent their support helped create the ideological premise of a 'democracy' that could actively impede the participation of popular movements in political decision-making (Robinson 1996: 77).

In the specific historical context of the collapse of military dictatorships in most of Latin America, the emerging model stressed the need to transform elites throughout the region. This new model summoned the Latin American nations to shape their re-emerging civil societies in the likeness of the political institutions of the United States (Alzugaray 2004). In support of these goals, the National Endowment for Democracy (NED) was created in 1983, officially termed a "private, non-profit organization to strengthen democratic institutions around the world through nongovernmental efforts ... governed by an independent, nonpartisan board of directors" (NED 2008). The NED was in reality a state-funded institution designed to promote the kind of 'democratization' that Washington saw fit, with its main emphasis placed on working with local private sectors and transnational capital (Robinson 1996).

The accelerating process of global capitalist expansion was by that time pressuring Washington to carve out new markets while securing its access to strategic raw materials. The ideological cover for this imperative could be found in the US posture as self-appointed 'guardian of the free world' and the form that it took was to promote 'democratic elitism' across the Americas. This meant that post-dictatorial civil societies in transition throughout the hemisphere were once more in the sites of powerful financial interests. At the same time, the new wave of US influence could count on an ideological apparatus that wielded unprecedented power, namely, a highly transnationalized mass media that projected the hegemon's worldview as universal. This ideological, economic and political offensive would work in smooth harmony with the Pentagon to ensure a new era of domination in Latin America.

The results could be seen throughout the hemisphere as electoral democracies became consolidated, bolstered by diverse forms of US interventionism and a whole host of 'punitive policies' that were deployed whenever needed to preclude the emergence of political alternatives. The new 'democracies' became increasingly bleached of progressive policies and stripped of their endogenous sensibilities. Particularly embittering has been the turn to the right of 'socialist' and leftist parties such as in Chile, Brazil, and elsewhere that transited through a social democracy phase before completing their metamorphoses to neoliberalism. The consolidated electoral regimes that proliferated through the region meant that genuine popular participation in decision-making was decisively off the table.

Genuine democracy, particularly in the context of underdevelopment, meant that citizens would need to be incorporated into the day to day construction of a socially inclusive path to development. This in turn implied that a democratic state could play the pivotal role of ensuring respect for human rights, protecting those social sectors at greatest risk, and presiding over a redistribution of wealth that could make democratic rule viable and effective. This kind of role was, of course, far from the concerns expressed by the new elite democracies that were being nurtured throughout the region. Vandana Shiva well captured the historic dynamic that was unfolding: "The combination of corporate globalization and electoral democracy is separating leaders and governments from society and people ... Globalization has pushed representative democracy to its final test" (Shiva 2005: 107). In those countries where the system eventually got strained past the breaking point, it became clear that a genuine path to democratization would likewise require building an alternative path to development.

Persistent Dependency and the State

Although the liberal tradition established restrictions on individual freedom to the extent that the state needed it to ensure its stable functioning, the growing need to regulate capitalist economies to deal with their recurring crises in time brought challenges. Contemporary democracies have been forced to alternate abruptly between extended periods of state intervention in the fiscal, regulatory, and redistributive spheres and periods of 'structural adjustment' where state involvement in regulation and service provisions is severely restricted. These cyclical swings have forced democracies inspired by the liberal tradition to stretch the boundaries of their ideology in order to legitimize the 'democratic' character of the state's changing requirements.

Since the erratic changes can best be thought of as a systemic re-accommodation to structurally induced dilemmas, we can identify two recent periods when these lurches were particularly acute in Latin America. The first was the shift to the 'developmental state' in the 1960s to the early 1980s, while the second was the imposition of the neoliberal state in the period that followed. In the first shift, the historical context revolved around post-World War II development initiatives that were attempting to systematically dismantle the surviving feudal structures that the region inherited from colonialism. Casted as a process of 'modernization,' a modest increase of state intervention in economic affairs was defined as strategic for accomplishing a rapid assimilation of 'superior' Western capitalist values. Here, the expansion of the capitalist state was presented as 'pragmatic' in recalibrating laissez-faire economies to the modern demands for new technologies and rapid infrastructural expansion. Furthermore, it was argued that modernization as a process required foreign resources for its developmental 'jump start,' managed by a rational-legal state that could overwhelm resistances from traditional sectors and accelerate the process of industrialization in preparation for an economic 'take-off.'

With historical hindsight, it is now clear that capitalist modernization amounted to a kind of 'imperialist decolonization' in accordance with the emerging needs of global capital expansion. The early forms of critical analysis began to coalesce by the 1960s after it became abundantly clear that modernizing approaches were deepening rather than eradicating the structural dependency that had earlier been constructed under colonialism. In the words of legendary German theorist Andre Gunder Frank (1966), capitalist modernization resulted in 'the development of underdevelopment.' The chorus of critiques grew (e.g. Dos Santos 1970; Cardoso 1972; Cueva 1974; Bambirra 1978; et al.) and gradually shaped the theoretical contours of what became the highly influential Dependency perspective.

As dependency deepened, so too did doubts about the governability of modernizing regimes. The most radical theorists of Dependency (e.g.: Ruy Mauro Marini; Theotonio Dos Santos) insisted more on a Marxist social class analysis and gravitated towards an anti-capitalist critique of modernization as a contemporary phase of imperialism (Sotelo Valencia 2014). Meanwhile, the more moderate voices within Dependency approaches explored the possibility of reforming capitalism and developing an endogenous model of development that could overcome underdevelopment by carving out a more autonomous (less dependent) strategy of national accumulation. This more prescriptive approach built upon earlier work of structuralist development theorists and ultimately proved more amenable to nationalistic reformers of the era.

The moderate and 'pragmatic' voice of dependency ultimately achieved its paradigmatic form in the regional think tank ECLAC (Economic Commission for Latin America and the Caribbean). This more nationalist and autonomist

alternative to capitalist modernization chimed with the earlier work of the
Argentine development theorist Raul Prebisch (1950). By 1950, the seminal
work of Prebisch based on his critical analysis of Argentina's declining fortunes
had already established the central premise of structuralist economics, namely,
that the pursuit of policies based on liberal comparative advantage theory only
led to a deepening dependency and decapitalization of developing countries.
The 'Prebisch Doctrine' thus established the futility of modernization and the
need for a state-centred alternative that could overcome the accumulating
syndrome of dependency. Upon his appointment as director of ECLAC, the
UN-affiliated research institution helped to catapult his ideas into worldwide
prominence that would become a magnet for policy formulators who feverishly
searched for the alternative policies that could break the bonds of structural
dependency.

ECLAC was therefore quite influential in helping to formulate the import-
substitution approach to development. Based on a Keynesian critique of liberal
economic premises and informed by the Prebisch Doctrine, it defended the
necessity for an endogenous-based industrialization strategy under the auspices of a 'strong state' that could take the reins of development and steer
it away from dependency and towards structural reforms aimed at a more
balanced growth. The strong state envisioned by this strategy would now be
the main agent of national development in systematically promoting agrarian
reform, regulating foreign commerce through selective applications of tariffs,
building an economic infrastructure based on fully national criteria, and even
engaging directly in the production of strategic goods and services as necessary to achieve these aims. The end result was a mixed-economy strategy for
autonomous development in which the state would employ whatever means
necessary to overwhelm whatever internal challenges remained to structural
reforms.

The insurmountable difficulty for the state in this approach to development
was to shed its own class character in order to confront competing fractions
of national elites, foreign interests and the social class demands of popular
organizations when they emerged as obstacles to the particular structural
reforms being imposed from above. Indeed, the region's earlier experience
with nationalistic reform strategies had been met with mixed results. The
experiences of Lázaro Cárdenas in Mexico (1934–1940), Peronism in Argentina
(especially 1946–1955), and the Estado Novo de Getulio Vargas in Brazil (1937–
1945) were all examples of populist regimes that posited the state as a defender
of the interests of the working class and as a reformer of the socio-cultural
context oriented by the demands of the popular sectors. In these cases, the
state played a key role in organizing the expansion of social and physical

infrastructure. The populist movement of Jorge Eliécer Gaitán in Colombia and the ephemeral populist government of João Goulart (1961–1964) in Brazil mounted similar approaches. But the populist 'strong state' never accorded a central role to popular participation. This left it equally compatible with military dictatorships of nationalist orientation, such as those that took power in Brazil, Peru and other countries during the sixties. The legitimacy of the state became reduced to the ability of military governments to 'deliver the economic goods.'

Neoliberalism and Its Contradictions

When global capitalist expansion was forced to deal with recurring recessions during the 1970s, a new strategy was deployed to extend its penetration into national markets. This brought with it a confrontation between the needs of transnational capital and those of unequal and underdeveloped Latin American formations. With few exceptions, the ensuing crisis in the ECLAC model of import substitution led to the imposition of a structural adjustment as dictated by the core economic superpowers. This resulting crisis within the corporatist and developmental state suggested that another wave of reform was imminent.

The new backdrop to development consisted of the emergence of a powerful conglomeration of immense transnational corporations termed the 'New Leviathans' by the Argentine theoretician Atilio Borón (2000). The sovereign national states of Latin America were now expected to accommodate the 'natural rights' of global capital. In retrospect, it is evident that the global restructuring of capitalist accumulation in the 1970s marked the beginning of what we now know as 'globalization' while the political and economic policies that supported it became known as neoliberalism. Under neoliberalism, support for a reformed state was considered synonymous with a promise of development for all, better living conditions and increasing levels of personal consumption. The instrument to achieve this would be the market, and a representative state would be relegated to assuring the passivity of civil society and the inviolability of private property rights casted as the best way to experience individual rights.

With few exceptions, such as the island nation of Cuba, neoliberalism became rapidly consolidated in Latin America. On the heels of two terms of neo-conservative rule under Ronald Reagan, a member of the George Bush (Sr.) successor administration promulgated what became known as the 'Washington Consensus,' i.e., a set of guiding principles for the new form of state rule that included: fiscal discipline; reordering public expenditure priorities; tax reform; liberalizing interest rates; a competitive exchange rate; trade liberalization; liberalization of inward foreign direct investment; privatization; deregulation; and the strengthening of all property rights (Williamson 2002). Indeed, the radical retreat from Keynesian principles became posited as the 'only' possible form of democratic development.

Of course, acceptance of Washington's dictates was not a peaceful and harmonious process. Neoliberalism was first imposed via experimentation at gunpoint in the case of Chile in 1973. By the early 1980s, badly indebted countries were subjected to structural adjustment as a condition of acquiring new credits for their survival. New initiatives from Washington worked to consolidate and cement into place these neoliberal policies through 'preferential' trade and investment policies that culminated in a whole set of sub-regional and bilateral Free Trade Agreements of which NAFTA was the first.

Neoliberalism sought to create its own mythology regarding the distribution of wealth, reviving the already discredited notion that capitalism is self-correcting and evolving towards greater equality. What were construed as inherent inequalities would remain at an 'acceptable' level, as the myth goes, and 'the magic of the market' would work to harness new and powerful technologies that would achieve the best possible world for the greatest number of citizens. Beneath its harmonious ideological mantras, the Washington Consensus sought to facilitate access to natural resources and cheap raw materials within the region, with a growing emphasis on energy. It further proposed trade policies designed to capitalize on US dominance over technology and information to deepen its hegemony over emerging Latin American economies. A new generation of global trade initiatives would seek to expropriate national states from their essential decision-making processes, placing them in the hands of non-democratic supranational institutions such as the International Financial Institutions (IFIs) and transnational corporations, and subjecting them to arbitration mechanisms that operate beyond the reach of national institutionality. In this scenario, since many transnational corporations had budgets much greater than those of most Latin American nations, the asymmetries would be considerably pronounced. In short, this amounted to a global capital offensive that aimed to enshrine the 'rights' of capital at the expense of the human and ecological rights of the peoples of the region and beyond.

The evident contradiction of this process was the intentional separation of economic decision-making from the mechanisms of democratic participation. Citizens who voted in elections were increasingly prevented from exercising their power over the market, which on the contrary responded to the dictates of large foreign corporations. Neoliberalism had a meticulous focus on dismantling all regulations and the interventionist capacities of national states, arguing that the 'free market' could not tolerate any form of 'state interference' in economic affairs. In the name of freedom, the economy would 'replace' politics. While Boron (2000) argues that in Western conceptions of democracy, since Plato, there had always been a certain concern for equality, neoliberalism aspired to achieve total indifference to the social distribution of wealth, thus creating a contradiction underlying the democratic pretensions of neoliberal politics. By eliminating the growing acceptance of basic rights in the name of economic progress, neoliberalism essentially expropriated entire social sectors from the effective exercise of citizenship.

It would not be until multiple crises erupted in 2008 that orthodox neoliberalism would come under attack by the global establishment in an attempt to re-stabilize capitalism. Most of the major Western powers in the context of financial crisis began to selectively warm up to neo-Keynesian sentiments, assigning to the state the role of the financial rescue of those very same entities that generated the crisis in the first place through their reckless speculation. That was only half of the picture, however, since instead of guaranteeing employment and a living wage to alleviate a chronic tendency towards overproduction and the contraction of aggregate demand, the attempt was made to have workers shoulder the weight of the bailout through wage cuts, the contraction of social spending, regressive fiscal policy, unemployment and repression (de la Barra 2009). This partial revival of Keynesian common sense kept the global economy in precarious straights while demonstrating how far the state had become removed from the democratic principles of representing the interests of the majority.

Latin America and Representative Democracy (That Represents No One)

In Latin America, democracies share a number of specific common factors. In several countries of the region, dictatorial governments were used under military regimes or authoritarian civilian governments to impose the dictates of US imperial hegemony. Since then, many of these countries remained trapped in a process of 'redemocratization' which itself was narrowly defined as a return to electoral politics. The contradictions of representative democracy in this context are manifold.

Many human rights violators of the preceding period still enjoy impunity and some still remain in power or have important advantages that allow them to be elected or placed in public office (Garretón 2011). The right to fully reveal the historical truths behind the savage authoritarian rule in several countries such as Chile, Guatemala and Argentina has remained largely frustrated. The preservation of impunity was one of the first principles mandated in allow-ing the return to electoral democracy (de la Barra 2007). Just as amnesties and generalized guarantees of protection for executioners and genocides were gen-erous, the preservation of anti-subversion laws aimed at the popular sectors proved to be ruthless in many Latin American countries. This was a testimony to the class base of the earlier phase of state terrorism and its underlying con-tinuity through the legal mechanisms of repression that persist in the return to civilian led democracies. The instruments remain in place for the use of state force in dismantling a wide range of social movements that have grown in protest against neoliberalism. This, in turn, limits the possibilities of organized and effective participation.

Corruption was another common thread that was wrapped around the return to electoral politics, thriving like never before in the absence of regu-lations that were mandated by the neoliberal model of development. A report by the Transparency International group concluded that by 2005, 23 % of Latin Americans thought that their governments actually encouraged corruption (Transparency International 2006). Political parties and the police were judged equally negatively, according to the survey data. Around one in three respon-dents who had contact with the police revealed that they ended up paying a bribe. Law enforcement mechanisms are perceived as plagued by corruption and the judiciary ranked as the third most corrupt institution.

Another study carried out in eight Latin American countries in 2007 by the Latin American Faculty of Social Sciences of Chile (FLACSO-Chile), found that both political parties and electoral finance lack transparency (Fuentes, Villar and Ríos 2007). The main problem identified was that political parties do not report on their private financing or make public reliable information on their overall financing. When politicians respond more to donors than to their constituents, it is the poor who suffer the most, given their inability to influence those who should represent their interests. Although funding for election campaigns varies from country to country, it is an almost universal source of corruption, allowing large-scale capital to exert a decisive influence on national politics. The affected citizens have no reliable way to know what economic interests their candidates support. The correlation between money and politics, as currently practiced in representative democracies at national and local scales, is the very negation of the three basic principles underlying any fair electoral process: (1) equality in the access of voters to the information regarding the different candidacies as well as the equality of opportunities of the candidates to express their points of view to the voters; (2) freedom of opportunity to become candidates, as well as equality of means to run for public office; and (3) the ability of voters to exercise their electoral option free of pressure or manipulation (Fuentes, Villar and Ríos 2007). Most funding for electoral campaigns comes from contributions from very wealthy individuals and large corporations. The tax exemptions granted are an important way of maintaining the antidemocratic relationship between money and politics. These fiscal privileges directly transfer the costs of the electoral campaigns to the citizens since the public budget is affected by the loss of those revenues. In addition, candidates selected through contributions from the private sector must defend the interests of those who have funded them.

A significant point where the crisis of electoral democracy is centred is related to the political parties in the region. For a good part of Latin American history, political parties have served as important agents of social change. At the beginning of the 21st Century, however, political parties throughout the Latin American political spectrum fell into a generalized crisis. When Mexico's Institutional Revolutionary Party (PRI) lost its long-standing monopoly on power in July 2000, it became a symbol of the collapse of many of the region's former political parties, particularly those associated with the 'strong states' of the developmentalist tradition.

The factors that led to the electoral collapse of powerful parties like the PRI are still the subject of analysis. Many argue that the layers of corruption accumulated in such traditional parties, the fiscal mismanagement by Latin American officials and a generalized crisis of legitimacy of representative politics were some of the factors that combined to push things to the breaking point. However, the crisis of political parties also had deeper, structural roots in the systemic (and anti-systemic) dynamics that began to converge in the 1970s, forcing parties like the PRI to switch to 'reformist' positions to the extent where they effectively undermined their own political bases of support.

As political parties evolve within a national context, they must continually struggle with specific social factors that delimit the nature of their organizations, their institutional reach, and the extent to which they are able to consolidate their political power. Analysts of dominant currents, close to Max Weber's tradition, tend to emphasize organizational aspects in their explanations of this process, such as the relative professionalization of political cadres (Colburn 2002), the bureaucratization of political parties and other dynamics of increasing 'rationalization.' More critical theorists opt for a Marxist approach that focuses on the inevitable tensions that exist between social classes within a larger dialectic of the developing economy in a global political economy context. This latter approach suggests the importance of using an historical approach to the topic when analysing the composition, decomposition and recomposition of political parties. It seems relatively clear that the dissolution of the specific social bases of support of many formally powerful Latin American political parties, structurally induced, corresponded to the accumulation crisis that affected endogenous development strategies during the 1980s (Dello Buono 2007).

The rise of neoliberalism under the Washington Consensus required the unbridled deregulation of economies and the transnationalization of regional economic flows. The neoliberalization of dependent capitalism throughout the region imposed complex changes in national decision-making structures. Once national economies were forced to open with the rapid liberalization of trade, many traditional political parties began to feel expressions of opposition among their usual bases of support. The pressures exerted on the parties to deepen the process of neoliberal reforms altered the overall political and ideological context. The consequent struggle to define themselves, along with the abrupt change of their political platforms, quickly threatened their very survival.

Although the consolidation of neoliberalism continued to mistreat the grassroots bases on which many political parties were erected, the sustained movement towards privatization of public services and the 'de-nationalization agendas' made it gradually impossible to maintain state levels of essential social spending. The privatization and financial transnationalization of the production of raw materials intensified the denationalization of the economic incomes that had been so strategic for development as formerly directed by the state. The high unemployment and the intense pressure to make existing labour regimes more flexible struck directly at the vitality of the unions in the region. The mutilation of the social spending base of national states coincided with the rapid growth of poverty and social exclusion, producing a crisis of legitimacy in the region that affected many parties in power. The pressure was also by political parties in opposition whose electoral strength was based on the representation and defence of sectors that were being adversely affected by neoliberal policies.

In some places, the situation became extremely volatile and explosive. The clearest example of this occurred in Venezuela when, in 1989, there were riots against the International Monetary Fund (IMF), known as the *Caracazo*.[1] This

---

1   The *Caracazo* in 1989 was the result of the neoliberal policies of Carlos Andrés Pérez that sparked an urban uprising. The violent suppression of the rebellion led to at least 400 deaths at the hands of the Venezuelan armed forces, this according to official sources (Human Rights Watch 1994). Among the popular sectors, estimates range to up to ten times as many, and several mass graves discovered later seem to vindicate some of their allegations. The *Caracazo* marked history in more than one sense because it helped consolidate a dissident movement of young cadres within the Venezuelan armed forces known as the Revolutionary

event was the first to reveal the resurgence of popular resistance that had been accumulating outside of the discredited political parties and the domination by elites. The revolt was violently suppressed while the political class of the region took detailed note of the facts.

It is clear that the use of state and paramilitary repression and the crisis of political parties in Latin America went hand in hand. During the previous reign of the military dictatorships, the limits placed on reformist populist parties that had successfully acceded to power were established by armed violence. As the limited social reach of these parties in power during the 1960s and 1970s became evident, more radical leftist parties gained strength. The violent response of the most reactionary forces of the continent fell directly upon them and this repression was solidly backed (and to some extent organized) by Washington and other foreign powers that acted as guarantors for elite rule. The wave of repressive 'dirty wars' led by the military class of the region was singularly dedicated to uprooting and eradicating all those forces of the left that had managed to establish popular bases. These repressive campaigns eventually delimited the small political space selectively granted to civil political parties.

During the subsequent period of 'redemocratization,' the Armed Forces of the region largely returned to their barracks, but continued to cast a long shadow over the political landscape. In several notable cases, some of the most abject human rights violators in the region retained their status as important political actors. In Argentina, for example, the collapse of the military dictatorship in 1983 led some members of the Armed Forces to form the *Carapintada* ('painted faces') movement that organized several rebellions against the civil government, demanding amnesty for the atrocities committed during the period of military rule between 1976 and 1983. When assuming office, the administration of Carlos Menem yielded to its demands in exchange for a quick suppression of the uprisings, declaring amnesty for previous dirty war crimes and allowing the *Carapintada* to assemble in the form of a new political party that allowed them to extend their influence over the entire political right (Payne 2000).

In Chile, the return to electoral democracy took place under the dictates of the Constitution formulated by the military government. This not only ensured impunity, but also appointed ex-dictator Pinochet as Senator for life and guaranteed an even greater share of public funds for the already well-funded Chilean military. In the case of Guatemala, former members of the

---

Bolivarian Movement-200 (MBR-200) that later resulted in the Bolivarian Revolution under the leadership of Hugo Chávez.

paramilitary forces known as 'Civil Action Patrols' (PAC) mobilized in opposition to the Peace Accords and to demand payments for their services during
their 'dirty war' against a Mayan and peasant insurgency. In this, they enjoyed
the protection of former dictator Efraín Ríos Montt who obtained immunity by
presiding over the national parliament. The ex-PAC soon emerged as the 'shock
troops' of the former dictator's political party, a model that was also established
in El Salvador, with paramilitaries serving the Nationalist Republican Alliance
(ARENA) party. In each of these cases, there has been a deep social polariza-
tion around the impunity of past crimes and the permanent role of the army in
civilian government, effectively creating an additional element of 'distortion'
in the political terrain of the parties.

Meanwhile, the 'war on illegal narcotics' launched by the United States
beginning in 1980 further worked to favour repression and militarization, particularly in the Andes and the Caribbean Basin. Coinciding with the establishment of neoliberal policies, Washington assumed a quintessential contradictory position. On the one hand, states were pressured to ramp up efforts to
combat illegal drugs by prohibiting and eradicating their cultivation and by
applying strict financial regulations. On the other hand, the neoliberal model
relentlessly insisted on a minimalist state and a laissez-faire stance. The fact
is that the illegal narcotics industry consists of immense globalized commodity chains, where production, consumption, trafficking and laundering of sales
share direct links to arms trafficking, corruption and, in many cases, electoral
politics. Ultimately, the inextricable connection between legal and illegal markets made 'narco-capital' a major economic force in the region. In Colombia, it
became the determinant force in key areas of national development. Likewise,
it would be difficult to imagine the current development of Panama without
taking into account the role of money laundering. Liberalization of the banking
policies facilitated the expansion of money laundering while the intense pressure of the neoliberal free trade agreements and the consequent cheapening
of imports led to the collapse of many traditional crops. This was further intensified by a continuing policy of US subsidies for their own domestic producers
while it sought to punish other countries for doing the same. The sum total
of these policies helped reinforce the trend of peasants turning to the production of illicit crops that form the basis for illegal trafficking in narcotics. In the
final analysis, the neoliberal formula to reduce the state's presence in national
finances contributed to generating the most favourable conditions imaginable
for the illegal traffic of narcotics and money laundering.

The magnitude of the lucrative activities associated with drug trafficking,
with collateral links to arms trafficking and financial negotiation, helps to
explain how regional politics became so easily corrupted. Political figures

became accomplices to multiple instances of corruption that attempted to shape political decision-making, effectively altering the exercise and config- uration of political power. The Andean and Caribbean Basin countries are clearly among the most dramatically affected by the dynamics surrounding drug trafficking, as evidenced by high levels of violence, corruption scandals and repression. The United States itself was certainly no exception since it became directly involved in drug trafficking during the 1980s to circumvent congressional bans and to finance the war against the Nicaraguan 'Contras.'

In what some have called the 'Colombianization' of the region, the US war on drugs gradually expanded, allowing the emergence in the 1990s of Plan Colombia as the prototype of a new interventionism. Plan Colombia, under the administrations of Bill Clinton and George W. Bush, had a dramatic impact on the political landscape of that South American country, forcing unpopular policies of aerial fumigation during successive Colombian administrations and even the cancellation of visas to some Colombian officials because of their ties to the drug cartels. These and other US interventions helped destroy the political perspectives of Colombia's two main parties and favoured the rise to power of a ferociously pro-American political force under the leadership of Alvaro Uribe, closely linked with Colombian paramilitary groups and drug lords. Similarly, anti-narcotic interventionism led to the fall of central figures in the Caribbean region and, in the case of Panama, to a large-scale military invasion that culminated in the US capture of its former ally General Manuel Noriega, thus paving the way for a shift towards neoliberalism by successive administrations.

What has changed the Latin American stage of political parties in a more hopeful direction has been the emergence of anti-neoliberal social movements. The abandonment of traditional political parties by the popular sectors con- tributed to a string of electoral victories of the left. In some cases, this has included the formation of new groupings of parties organically linked to the popular movements that serve as a political instrument to propose an eman- cipatory agenda. A successful example has been the Bolivian Movement for Socialism (MAS) led by Evo Morales, who in 2006 managed to form a majority government in power. A year later, Rafael Correa and his Alianza País coali- tion were victorious in Ecuador. In both cases, the electoral victories of the left mounted on the failure and discredit of traditional parties and the apparent inability of the traditional left to take the initiative.

In 2005, the *Frente Amplio* coalition, comprising more than a dozen leftist parties, including the former urban guerrilla Tupamaros who led the clandes- tine struggle against the Uruguayan military dictatorship (1973–1984), won the presidential elections in Uruguay. The victory was the product of an 'accumula-

tion of forces' that had begun with the founding of the Front in 1971 and culminated in the end of the domination of a bipartisan conservative system (Cores
2007). In the case of Brazil, the 2002 Workers' Party (PT) electoral triumph that
led Luiz Inácio da Silva ('Lula') to the presidency was built on successive successes at the state levels. Forged in the union struggle that formed the vanguard
of the popular resistance against the military dictatorship (1964–1985), the PT
was constituted in 1979 as an electoral party that offered workers an alternative
in the political field, a 'third lane' positioned between social democratic and
vanguard Leninist parties (Keck 1992).

In Venezuela, the electoral victory of Hugo Chavez and his Fifth Republic
Movement (MVR) in 1998 provided another spectacular blow to the domination
of traditional political parties. The MVR was formed as a political organization
in 1997, when Chavez began his campaign for the presidency. Later, in 2006,
Chavez proposed the dissolution of the MVR in favour of the formation of the
United Socialist Party of Venezuela (PSUV). The emphasis of this proposal was
on the creation of a self-sufficient unit of popular power that articulated with
the state. The idea was to organize a restructured party in contact with the
bases, which would select its leaders democratically.

Constructing Democracy with an Emancipatory Agenda

The role played by the state in development is a complex and highly contradictory one. In the most fundamental sense, a capitalist state actively maintains and reproduces the overall system of capital accumulation that requires
a nearly continuous involvement in the economy, regardless of the ideology in
place at the moment. A common example is when the state directly engages
in areas of production that are unprofitable for private capitalist enterprises
or which are determined to be too essential and/or sensitive to be entrusted
to private producers, such as essential social services, national defence, and so
on. Establishing the optimal parameters of state involvement in a developing
economy is ultimately a strategic part of the political process.

At the same time, the state must ensure that the overall process of capital
accumulation continues uninterrupted despite political disputes and social
conflict. To fulfil this role in a liberal-democratic capitalist context, the state
assumes the position of 'mediator' in which it organizes, regulates and protects
the 'legitimate' institutions of conflict resolution that have evolved in a given
national setting. By acting as mediator, the state actively maintains its image of
a 'neutral' governing body that legitimately establishes the rules of the game by
which all social classes must play.

This dual character of *reproduction* by intervention and *mediation* on the part of the state helps organize society in a manner that ensures the smoothest possible operation of a given political economy. By administering essential economic processes while adjudicating conflictual political dynamics, representative democracies operate in a manner that attempts to at least appear 'democratic.' The limits of a representative state's democratic character are challenged, however, when its policies become thwarted by political opposition, or when social struggles demand changes that conflict with the reproduction requirements of the existing political economy.

In the course of development, reproduction operates as a dynamic process that involves state utilization of the legal system as an instrument of regulation and social control. To analyse these dynamics, sociologists of law refer to the 'legal order' of society to include all those state institutions involved with the formulation, implementation and enforcement of laws that seek to define the normal and acceptable patterns of social conduct. The legal order in turn forms part of a larger set of relationships defined by the existing political economy in which they are located.

While the legal order of capitalist societies fundamentally operates as an instrument of class oppression, enforcing the class interests of the rich and powerful at the expense of poor and exploited classes, it does not always do so in a mechanical manner. In a representative democracy, the nature of the legal order is such that it must guarantee the continued accumulation of capital and profit, but at the same time help maintain a semblance of fairness in the system. In this sense, the legal order has a dual and conflictual nature that allows for legal 'victories' by non-dominant social groups in instances that do not challenge its essential structure and thereby provide the urgently needed semblance of legitimacy.

All of this becomes even more complex in a context of social transformation. When organized social resistance to an established order becomes so great as to challenge the existing arrangement of social class domination, a revolutionary situation emerges. In this kind of revolutionary situation, the state can be 'captured' by a new alliance of social classes and transformed into an agent of social change. If it proves capable of developing new forms of intervention in pursuit of an agenda of transformation, it can ride on the crest of a new configuration of ascendant social class forces that favour an emancipatory agenda.

An alternative scenario emerges when the state continues (or reverts back) to reproduce the previous social class order and its accumulated institutions. This actively impedes a transformative agenda through a variety of mechanisms that absorb, co-opt or repress existing political challenges to the established order. If the state moves in this direction, social transformation becomes virtually impossible, no matter how well-intentioned or ideologically committed to change that the leaders of a given government may be.

One of the greatest *limiting* factors of those political parties seeking to bring about emancipatory social change is the carefully designed legal regulation of the electoral arena and constitutional order. Capitalist representative democracies have gradually accumulated a complex array of legal mechanisms designed to resist fundamental change and establish the parameters of possible reforms that can 'fine tune' social institutions without disturbing their essential class character. This means that political parties seeking fundamental social change usually encounter numerous legal obstacles that impede their electoral prospects and limit the possible scope of changes, even when an electoral victory of radical leadership proves successful. For this reason, political parties who seek fundamental transformation have only just begun the process of orchestrating social change upon electoral victory. An essential component of implementing an emancipatory agenda ultimately rests on a successful *restructuring* of the legal order and this typically implies constitutional reform.

Social movements, as opposed to traditional political parties, tend to enjoy greater latitude in their visions, methods and proposals for fundamental social change as they typically remain more intimately connected with their social base. Equally important is the fact that they tend to be less 'filtered' through heavily structured legal mechanisms of the sort that regulate political party activities. By remaining to some extent 'outside' of the institutionalized channels of a given society, social movements remain uniquely positioned to develop alternative understandings and beliefs that call into question the 'official definitions' put forward by the state. Therefore, they help to reveal underlying injustices or other shortcomings of 'progressive-minded' governments that are either excessively bureaucratized or politically compromised.

Due to the organizational nature of most liberal representative democracies, social movements remain largely deprived of mechanisms to directly affect the established legal order and attendant monopoly over the exercise of legitimate force wielded by the state. As a result, social movements quickly come up against the legally established limits placed on social protest. To the extent that they move outside of these legally recognized channels, they are criminalized and face the full fury of state repression. Criminalization of their activities also challenges the legitimacy of social movements since this seeks to stigmatize movement members as 'socially undesirable' and as dangerous 'criminals.' For these and other reasons, social movements left on their own frequently tend to remain trapped by the established legal constraints governing social protest and are unable to bring fundamental social change to fruition.

If an emancipatory agenda is seriously contemplated via an electoral strategy, it becomes indispensable that the electoral victory of a progressive political force enjoys sufficient mass support by organized social movements. This is essential to support a process of fundamental change in the established legal order as part of a progressive transformative agenda. The process of transforming the constitutional framework of national systems of legality is one fraught with risks. It is a complex process of playing according to the established legal rules of the game in order to re-write those very same rules, working within the lawfully established processes for reforming or restoring a national constitution. It points to crucial issues such as which social forces control a constitutional assembly and to what extent entrenched elites continue to hold sway over the process.

In the context of an abrupt political shift in political orientation of an elected government, constitutional reform will almost invariably inspire conflict. Any misstep along the way can result in the eruption of violence and possibly an extra-legal intervention of the armed forces, either at the behest of historically entrenched national elites or on the part of foreign powers that stand to lose out from the unfolding process, typically acting in defence of transnational corporate interests or to address supposed national security threats. As history has demonstrated, the devastating results can include the imposition of a military dictatorship, foreign occupation and/or other scenarios for expropriating popular sovereignty.

Even if successful, there is never a guarantee that a progressive legal order will prove to be an effective instrument for change unless there is adequate political will to comply with it. An abrupt political shift in orientation also implies a need for the right kind of public policies and budgetary allocations to enable compliance. If on the other hand there is no overarching revolutionary process, it is very difficult to promote positive change regardless of whether there is a body of law that will permit it. The peaceful path to transforming constitutional frameworks throughout the region in order to enable forward movement in tandem with progressive political upsurges certainly entails a multitude of risks. In this scenario, the strength and concerted action of social movements in the struggle for a post-neoliberal future is indispensable both to the process of carrying out far-reaching legal changes and in subsequent efforts to ensure compliance with a more favourable legal order once established.

Conclusion

Understanding democracy as simply the holding of 'free' elections is akin to understanding economic prosperity as the achievement of 'free' markets. In the real world, neither of these institutions can operate 'freely,' but even if they could, they still wouldn't suffice for the desired goal. The creation of electoral representative systems was nevertheless proclaimed by the 'Washington Consensus' to be the *sine qua non* of democracy, allowing it to claim a semblance of legitimate democratic aspiration while promoting maximum opportunities for the domination of economic interests both inside and outside the countries of Latin America. Indeed, Washington's strategic plan of domination rested upon the structural fragmentation of the region's traditional political foundations and the consolidation of new regimes favourable to its hegemonic interests.

But despite Washington's best efforts, the limitations of representative democracy have not gone without notice by the popular sectors of Latin America who have been on the receiving end of neoliberal regimes. Social movements have increasingly mobilized to confront the failure of neoliberal development strategies to provide for sustainable improvements in the standard of living for the majority of Latin American peoples. This social resistance has likewise tended to unfold in an uneven manner but continues to demonstrate great potential despite the fact that its results have proven highly variable across the region. Its greatest successes have been felt in the creation of broad opposition fronts against neoliberal policies. The brakes were slammed on privatization in many countries and demands for greater social spending placed the social debt on the national policy-making table for greater scrutiny. In some cases, this has dovetailed with successful electoral strategies of anti-neoliberal candidates from non-traditional, leftist political formations.

Rather than simply changing the actors of weakened states, it is necessary for national states to regain their sovereignty and to work towards a regional sovereignty that allows genuine development in the interests of the popular majority. In this light, democratization means exactly the opposite of what neoliberalism has postulated. There is a need to rebuild the capacity of states to intervene in economic and social life with a view to reducing dependency and unequal development. In the most radicalized countries such as Venezuela, Bolivia and Ecuador, the construction of more participatory democracies was placed on the national agenda. The larger objective is to develop proposals that defend viable strategies for social transformation that go well beyond electoral institutions with a view to achieving a more genuine democracy that reintegrates popular decision-making with development.

For any of these processes to flourish, it is first necessary to identify and dismantle the traditional legal and constitutional mechanisms established over time by entrenched elites to prevent fundamental social change. The need exists to reformulate the rules of the game in favour of a transitional order in which continual, fundamental change can be protected from derailment by oppositional forces wedded to the status quo. The supra-constitutionality of constitutional assemblies is a particularly crucial issue if far reaching change is to be made possible. Otherwise, the existing legal order will protect elite interests and preclude any meaningful institutional change.

Elections for constituent assemblies where everybody, by definition, is a potential candidate are definitively more democratic than traditional parliamentary elections. Their participatory character generally favours a constitutional outcome that displays greater inclusiveness and popular incorporation into decision-making. In addition, constitutional referendums will have the ability to continually consolidate democratic constitutional processes. The challenge is to shape a legal structure that will enable continual transformations that can benefit the popular sectors while simultaneously opening the space necessary for mass participation in the governing process. A determinant factor in the success of these reforms is whether there are sufficiently strong social movements actively supporting the change process. If there are not, constitutional change will almost certainly be unsustainable, ineffective and highly susceptible to reversal.

Ironically, as soon as profound democratization begins to be achieved, new democratic mechanisms can provide spectacular opportunities for conservative forces to remount a legal challenge to transitional processes. Reactionary states under control of neoliberal-aligned elites have demonstrated their preference for repression and the criminalization of social protest, including collaboration with national armed forces. This has made resistance a very risky business and underlines the contradictory nature of a sustainable process of deep democratization and social emancipation, reiterating the high priority that must be placed on popular unity and concerted social mobilization.

It is possible that the region is poised for significant new advances that can better integrate new forms of democratic participation with economic and social development in the interests of the popular sectors. Some movement may be promoted 'from above' by states where radicalized political currents have come to power, while additional movement may come 'from below' by mobilized elements of the region's civil society. The path is likely to be fraught with risks since history does not now nor has it ever advanced in a simple linear fashion. What is clear is that the declining lack of legitimacy in elite-ruled representative democracies has intensified under consolidated neoliberal rule. The importance of social mobilization in driving democratizing and emancipatory processes forward and making them sustainable cannot be sufficiently underscored.

CHAPTER 8

Acquiring Technology in the Mexican Private Sector: A Disarticulated 'Linkage' of the Triple Helix

Miguel Omar Muñoz Domínguez

Our tendency to import theories that circulate in the first world, many times in an uncritical way and without the material sustenance they presuppose, is well known. Our approach to the production of knowledge and technology constitutes no exception. In this study, our intention is to contrast a prominent model of organizational cooperation as formulated abroad, namely, the Triple Helix, with the actual way in which the technology used in production processes by Mexican business is obtained. In developed capitalism, the Triple Helix refers to the union of the government, institutions of higher education, and the business sector, all working in favour of the generation of technology and innovation for production and economic development.

Although this model has been highly publicized and promoted for the policies of science and technology in our country, the prevailing method of acquiring innovation has been through its purchase. This is due to the formation of an economic ecosystem in which the developed countries technologically colonize underdeveloped economies, establishing a relationship of dependence where the actors involved in the processes of generating science and technology actually contribute to maintaining this underdevelopment both in their manner of thinking and acting. An effort will be made here to empirically verify this argument through existing surveys regarding the links by which Mexican entrepreneurs satisfy their technological needs.

The Model of the Triple Helix

The Triple Helix Model as proposed by Etzkowitz and Leydesdorff (1997) has had a marked influence from the late twentieth century to the present. The importance of the link between the university and industry lies in its potential to achieve the intended economic development much desired by many nations. In developed countries, relations between academia and industry are a key factor in maintaining their privileged status in which knowledge plays a

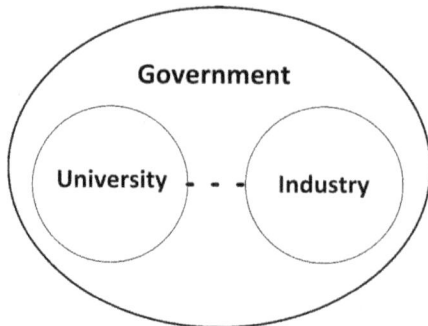

FIGURE 8.1 *Static model*

central role. Hence, the knowledge produced in universities is transferred in the form of technology to industry, intensifying a close relationship between these two spheres.

The model goes beyond establishing a superficial relationship between university, industry and government. Indeed, it implies an internal transformation within each of the spheres while the interaction between each involves further mutual transformation (Chang Castillo 2010). Henry Etzkowitz (2008) describes paradigms on the relations between the participating spheres that pave an evolutionary pathway to reach the ideal represented in the model. In the first instance, a static model describes where government controls academia and industry (see Figure 8.1). This case applies to countries where the government is a dominant institution and plays the role of coordinator where it leads and develops major projects as well as the promotion of resources and new initiatives. Here, industry and academia are posited as relatively weak institutional spheres that require state guidance. The former Soviet Union, socialist countries and, in a more diluted version, Latin American countries exemplify this static model of societal organization. When this model is translated into policies on science and technology it is characterized by universities dedicated to teaching which remain distant from industry.

In the second, 'laissez-faire' model (see Figure 8.2), the principal components interact only in a very modest way due to shared boundaries that are still sharply delineated.

Finally, there is another view (Figure 8.3) where the spheres of government, university and industry are interspersed "by generating an infrastructure of knowledge in terms of a superposition of the institutional spheres, where each one takes on the role of others and with emergent hybrid organizations in the interfaces" (Chang Castillo 2010: 88).

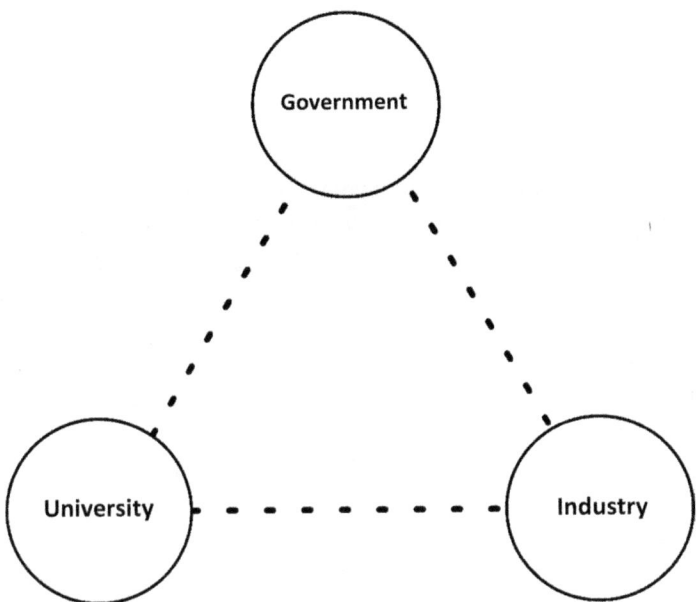

FIGURE 8.2 *Laissez-faire model*

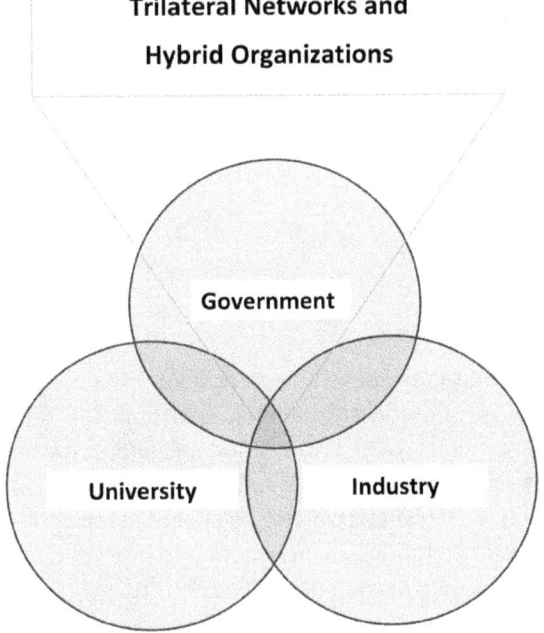

FIGURE 8.3 *Model of the triple helix*

In theory, the institutional and inter-institutional framework that "ensures the proper collaboration of the different spheres of which the generation and timely use of innovation depends [should be constituted] by norms, organizational cultures, perceptions, networks, information systems, capacities for leadership and implementation, among other factors, which together determine the capacities to develop activities of linkage" (Cabrero et al. 2011: 88).

The initial impetus for this model can be traced back to American capitalism of the first half of the Twentieth Century. It was in the context of World War II when "the elements of direct collaboration between university, industry, and government were invented [...] and military research links were maintained in academia" (Casas and Luna 1994: 3). Following the war, basic research came to be financed by government funding and during the 1970s research groups began to adopt organizational characteristics of private firms.

At the end of that decade, the US government implemented an indirect industrial policy for universities which implied that those who accepted federal research funding should capitalize on the knowledge they produced and transfer it in the form of technology to industry, making them active agents of technology transfer. Etzkowitz (1993) considered that a change was taking place, oriented "toward a multipolar interaction in which governmental authorities at different levels (international, national and regional) are relevant actors even in capitalist countries. A new mode of production is emerging in which academia, industry, and the state are no longer separate" (Casas and Luna 1994: 3). It can be assumed that in general, most developed countries have followed this sequence, including the establishment of scientific parks, the granting of venture capital financing, the generation of flexible local human resources, all framed within a specific economic moment that embraced the neoliberal model.

General Labour and Development

Victor Figueroa (1996) considers that knowledge and its development constitute a *condition* of the direct process of production and its expansion. This reiterates the Marxist categories of 'general labour' and 'immediate labour,' where general labour is understood as signifying scientific, intellectual, technological creation, while immediate labour is that labour which is in charge of the application of the fruits of scientific and technological advances in the productive process. The basic notion of this reasoning is that developed countries do both types of labour while underdeveloped ones essentially only carry out immediate work. In order to advance their process of accumulation, underdeveloped nations perpetually find themselves in need of buying foreign general labour. In this manner, "the development of the productive forces is located abroad, so that industry in underdeveloped countries is annexed to production in developed countries. This is what we term industrial colonialism" (Figueroa 2001: 11).

In a developed country, large industry achieves the linkage of universities and private enterprise in an institutional sense, just as depicted in the Triple Helix model. This linkage is seen to occur in the most optimal way so as to harness the general labour necessary for the development of the productive forces. So in places "where general labour is not exploited, there is not only the absence of such a link, but the absence of development itself in accordance with the needs of the era" (Figueroa 1996: 57).

The Concept of Linkage

Guillermo Campos Ríos and Germán Sánchez Daza (2006) discuss the difficulty of establishing a concrete definition of the concept of linkage, analysing the common sense interpretation that confuses it with other activities such as the extension and provision of services. They distinguish three types of linkage:

a) The 'economistic' approach, associated with obtaining royalties, in order that "through the sale of university products and services, 'juicy' results will be obtained in terms of economic resources for universities" (Campos Ríos and Sánchez Daza 2006: 22). This economistic vision usually rises to prominence in times of budget cuts, with the idea of using it as a palliative in the face of the financial crises in universities. However:

> The empirical evidence indicates that, even in the cases of the most developed universities in our country [...] in which the link with the industrial and productive sector has crystallized through important agreements and advisory contracts, the resources that it actually contributes still represent small percentage amounts with respect to their total budgets.
>
> CAMPOS RÍOS and SÁNCHEZ DAZA 2005: 22

b) The 'fiscalist' approach, associated with the understanding that the institutions of higher education (IHE) must perform certain social welfare functions with society, where almost anything is susceptible to being recognized as a linkage. This type of approach becomes confused with extension and university outreach.

c) The 'new function' approach to the university, under which the linkage is added to the traditional substantive functions of the university, i.e., teaching, research and extension. In this approach, 'the new function' is considered to be a structuring axis of academic planning through which it can establish:

> [...] a new social contract between academia and society. This requires broad and strong government support, in accordance with the role that has been assigned to research in the new economic model. The adoption of this new contract and its translation into implementation will obviously vary from one institution to another and will depend to a large extent on the response and support of national and international policies.
>
> CAMPOS RÍOS and SÁNCHEZ DAZA 2006: 23

Similarly, Luis Malagón Plata (2007) affirms that the pertinence of higher education has now become closely tied to the concept of linkage. He considers that linkage is understood through three fundamental notions: a) *responsibility*, conceived of as the appropriation and understanding of the social responsibility of higher education and its obligation to inform, explain, justify and respond to the larger society about the use of resources; b) *relations of community trust* regarding the participation of universities with the broader community in the development of higher education; and c) *market links* related to the sale of goods and services to industry, commerce and to those who need their products.

This last form of economistic linkage, together with the Triple Helix Model, has had many followers in Mexican government administrations. The idea of linking institutions of higher education with the private sector in order to positively achieve development dates back to the mid-1980s, which generated subsequent changes in both educational and scientific-technological policies. Even if there has been some progress in this area, it is not the paradigm that has become widely generalized nor has it managed to achieve outstanding figures (Rivera Vargas 2011).

The Disarticulation of the Technological Process

A program of production must have certain characteristics in order to effectively respond to productive activity. The means of production have the distinctive characteristic of being 'updatable' through technology. Machinery, means of transportation, tools, custom equipment, and software, among other elements, define productive capacities through that essential characteristic that gives them their degree of modernity (García Echevarría 1993).

Technology is not neutral in its economic sense. The ownership of the technology associated with its degree of modernity is linked to the nature and evolution of the development processes of different countries as inserted in an international economy. In the same territory, it can also present a disparity of the types of economic enterprise, and thus display functionally distinct processes of accumulation. These aspects demonstrate the diversity of interests at stake and "constitute the bases for the emergence of a set of social conflicts, the epicentre of which resides in the technological problem, which if it cannot find an adequate solution within the scope of the state, it translates into the disarticulation of the process" (Barsky 1980).

For Osvaldo Barsky, the disarticulation of an organization or system causes a vacuum in its operation as it lacks the necessary functional connections. He presents several models of disarticulation, for example, when the economic policy implemented by the state is not consistent with a rapid technological progress through the intensive use of the factors of production, or when "due to certain conditions that constitute a particular state constitution, it expresses a disinterested opposition on the part of dominant social groups in the technological process itself" (Barsky 1980: 21).

The Idiosyncrasy of Business, Academia and Government

The linking model of academia and private enterprise as we have seen comes to depend on a state policy and the economic position that an economy plays within global capitalism. It also responds to the ideological baggage of its actors, something which itself is a product of historical, cultural, social and economic processes being experienced by a country. Thus, in an economy characterized by innovation, the entrepreneur is an active actor who is responsible for being the catalyst. In this last regard, Joseph Schumpeter would emphasize that: "The creative action of the entrepreneur would be the motor of economic progress. This is manifested through the introduction of 'innovations' into the production process" (Furtado 1974: 48).

According to Dale Story (1983), industrial development in Latin America has been inhibited by national and international structural factors. Among the factors that render a domestic impact are the ideological limitations and political activities of Latin American industrial elites. Heriberta Castaños-Lomnitz (1997) carried out a study based on interviews with a selected group of 44 Mexican leaders from industry, government and the National Autonomous University of Mexico (UNAM) around the general theme of modernization, industrialization and technology transfer in the context of higher education. The representatives of all three sectors expressed a discourse of empathy with modernization and with the fact that higher education should play a more active role in the transfer of modern technology to Mexican industry. However, once conversations became less formal and were conducted in an environment of greater trust, informants from all sectors tended to express a resistance to technology transfer albeit for diverse reasons.

By dividing informants into three groups, namely, entrepreneurs, academics and public officials, Castaños-Lomnitz uncovered that business offered a discourse that was "emphatic in rejecting the idea that universities should focus on technology transfer to industry. Their job is to educate competent and well-behaved employees for jobs in industry. Public universities, in particular, are not doing their job at this point" (1997: 370). For example, one entrepreneur who happened to be a former UNAM research scientist offered the following opinion:

> What is the use of the university? Academic standards are low. People graduate with a bare minimum of know-how. Tax money is wasted on an academic apparatus which ejects graduates into the marketplace without considering the needs of society. It is just an arena for politics. Mexican industry imports the equipment from abroad and all we do here is maintenance [...] We can talk about university–industry relations some other time.
>
> CASTAÑOS-LOMNITZ 1997: 370–371

Another entrepreneur was of the opinion that the issue of discourse on technology transfer on the part of the UNAM is only 'rhetoric' and is probably motivated by political reasons: "Now the universities claim they want to peddle technological 'solutions' to industry. The socialistic intellectuals who ruled Mexico for decades corrupted the universities and first and foremost, UNAM. Even the engineering graduates are tainted with Marxism" (Castaños-Lomnitz 1997: 371).

A more recent survey on linkage that analysed the Political Science Department at the Autonomous University of Zacatecas (UAZ) gathered similar business opinions, both against institutions of higher learning as well as existing government policies. Take the following two examples of entrepreneurial opinion:

1)   Our country has not had governing officials who are concerned about the
     development of industry and technology, so we remain a country that

consumes this type of productive inputs. For example, it is enough to just mention PEMEX (Mexican Petroleum S.A. de C.V.); and

2)   I realize that there is complete disconnection of the universities for development of what they do or create. I have always said that there is a fertile field outside of the university institutions that they themselves do not know of. It's good that you are doing this survey. I'm micro and I need help, but I doubt that I can directly find it because they have not developed in a way to do this. They have no experience in order to take others by the hand (UACP-UAZ 2014).

Angel Plastino theorizes about the prominence of this type of attitude:

> In no time, anywhere except in Latin America, has it been believed that the mission of the university can be reduced to the mere training of professionals. Moreover, nowhere in the so-called First World are there institutions that do not develop active research and extension work and still be considered universities. This 'conceptual degradation' is a typical creation of the Latin American oligarchies.
>
> PLASTINO 1993: 46

Regarding academic opinion, the comments similarly go against a real link for technology transfer:

> The industrialists are rich, and in fact, they are mostly foreign. Should we rescue them? They have the luxury to be able to afford to pay for all the technology they need. They prefer to use foreign technology anyway. If the bridges and roads could be imported from abroad, where do you think Mexican engineering would be today? [...] Mexican industry is so backward that the very idea that the UNAM might be able to make a difference is absurd. We are not very knowledgeable about their needs and they [the industrialists] are not able to tell us. They do not even know each other, [...] the relevance [of our university] cannot be determined when trying to connect the UNAM to industry by force. Our university was not created for that.
>
> CASTAÑOS-LOMNITZ 1997: 372

Finally, within the discourse of the federal government, Castaños-Lomnitz's survey found opinions that were seemingly conciliatory but in reality pessimistic as suggested in the following:

We Mexicans have to get a move on it. Industry needs the university and vice versa, but neither one of the two knows it. Surely universities and industry can learn to cooperate. The question is under what conditions? Universities still seem to think in terms of government grants for joint technology projects with industry. That is out. We have tried it and it doesn't work [...] the UNAM itself is to blame for its situation. Our university wants to deliver to industry, but it cannot get its own departments and faculties to work together [...] the universities need money and applied research sells. Good for them, if they have something to sell. But that is what the market decides. The development banks could lend a hand [...] making universities provide technology for industry is unrealistic, [... as] industry evolves with a rapidly changing market view, and universities cannot keep up with it [...] university-industry relations will not occur unless the whole system is reformed and that includes the state.

CASTAÑOS-LOMNITZ 1997: 373

One gets the sense that government officials do not have a clear idea of policy design that could meaningfully achieve the goal of technology transfer between universities and enterprises. In sum, Castaños-Lomnitz (1997) identified a conflict between what she calls a pre-industrial discourse and a technological discourse. The latter is the one shown to be observable in the public sphere, presently enjoying a marked prestige. In reality, it seems that the skills provided by higher education are not intended to be relevant to industrialization. In this sense, it could be deduced that the modern university and business discourses coincide. Since students are not focused on the development of technology, they end up as modern foremen in the industrial sector, that is, they manage the production plants and provide maintenance for the technology being bought abroad.

Recently, José Franco, Coordinator General of the Scientific and Technological Consultative Forum (FCCyT) stated categorically that "the business and academic sectors are disconnected and have virtually no communication vessels. Technological innovation is one of the Achilles' heels of the Mexican economy because it has developed very poorly in our country" (Sánchez Jiménez 2014: 1).

The Lack of Endogenous General Labour Linked to Business in Mexico: ESV and ENAVES

In order to empirically demonstrate the lack of general labour domestically created by Mexican companies in connection with the institutions of higher education, the almost complete lack of hybrid organizations, as well as the actual way of obtaining technology that is being applied by the Mexican business sector in production, we will present the results of two surveys: the ENAVES (National Survey of Enterprise Linkage) and ESV (Survey on Linking the Department of Political Science).

The ENAVES was created in 2009 by the Ministry of Public Education and the Centre for Economic Research and Teaching. Its sample was taken from the SIEM (System of Mexican Business Information) and the 2004 Economic Census of the National Institute of Statistics, Geography and Informatics (INEGI). The ENAVES has a confidence level of 95% and a 3% margin of error. This survey provides its resulting databases in a format usable in the Statistical Package for the Social Sciences (SPSS) software.

The ESV was conducted between June and November of 2013 and its sample was taken from the business directory of the Mexican Business Information System (SIEM) of the Secretary of the Economy. It was applied to companies mainly from the industrial sector that registered valid and active emails through which they were sent a link to fill out an electronic questionnaire. Due to these conditions under which it was performed, the survey has a confidence level of 90% and an 8.7% margin of error, thus, the statements we can make from it have only an exploratory character.

According to ENAVES, 25% of companies have sought to link themselves with institutions of higher learning, but only 13.81% managed to implement this in practice with an aim of developing and disseminating knowledge. Incidentally, the corresponding figure for the United States is 50% and 45% for Brazil (González 2012: 1). This same survey showed that just over 70% of the existing linkage could be found in the manufacturing and service sectors, with 49.3% in services and 21.13% in manufacturing industry.

Specifically, the linkage for developing research and management was, according to ENAVES, 7.39% while for the acquisition of technology, the linkage was at 6.22%. Both of these rubrics can be broken down into 5 and 13 components, respectively. Within the first rubric, essential components of the Triple Helix were emphasized in the sense of institutional hybridization, such as joint research or participation in board directing. Table 8.1 shows the participation in this type of linkage by sector.

It can be observed that the manufacturing industry reaches barely 1% in joint research with IHE and that their joint participation in common board management does not reach 0.4%.

Similarly, it can be seen in Table 8.2 that the acquisition of technology by the manufacturing industry through purchase from IES is zero. The contracted research is a 0.39% and the creation of strategic alliances represent 0.58%.

TABLE 8.1   *Research and management of Mexican companies in cooperation with institutions of higher learning*

	Agriculture, livestock	Mining, electricity	Industrial manuf.	Trade	Transport	Services	Other
Joint research	0.00%	0.19%	0.97%	0.39%	0.00%	1.17%	0.39%
Participation in academic forums	0.00%	0.19%	1.17%	0.19%	0.00%	2.33%	0.78%
Participation in business forums	0.00%	0.19%	1.56%	0.58%	0.00%	2.72%	0.58%
Public-private forums	0.00%	0.19%	0.78%	0.39%	0.00%	1.95%	0.39%
Participation in administrative bodies	0.00%	0.19%	0.39%	0.39%	0.00%	1.17%	0.19%

SOURCE: DATA FROM ENAVES

TABLE 8.2   *Acquisition of technology by Mexican companies from institutions of higher education*

	Agric., livestock	Mining, electricity	Industrial manuf.	Trade	Transport	Services	Other
Licensing	0.00%	0.19%	0.00%	0.39%	0.00%	0.78%	0.19%
Purchase	0.00%	0.19%	0.00%	0.39%	0.00%	1.17%	0.19%
Technological innovation	0.00%	0.19%	0.78%	0.39%	0.00%	1.56%	0.19%
Technological assistance	0.00%	0.19%	1.17%	0.39%	0.00%	1.17%	0.19%
Patent rights	0.00%	0.00%	0.39%	0.19%	0.00%	0.39%	0.19%
Contracted research	0.00%	0.00%	0.39%	0.19%	0.00%	0.58%	0.00%
Rental of laboratories	0.00%	0.00%	0.00%	0.19%	0.00%	0.19%	0.00%
Business incubators	0.00%	0.00%	0.00%	0.19%	0.00%	1.17%	0.00%
Joint start-ups (joint venture or strategic alliances)	0.00%	0.00%	0.58%	0.39%	0.00%	0.97%	0.19%

	Agric., livestock	Mining, electricity	Industrial manuf.	Trade	Transport	Services	Other
Partnership in new companies	0.00%	0.19%	0.19%	0.39%	0.00%	0.97%	0.19%
Technological/ scientific parks	0.00%	0.00%	0.19%	0.19%	0.00%	0.39%	0.00%
Joint research centres	0.00%	0.00%	0.19%	0.19%	0.00%	0.58%	0.19%
Promotion of entrepreneurial culture	0.00%	0.19%	0.19%	0.39%	0.00%	1.36%	0.58%

SOURCE: DATA FROM ENAVES

The corresponding value regarding the existence of joint research centres is 0.19%. For the service sector, only slightly higher values were across the same rubrics. Null values stand out for the transport sector as well as the Agriculture and Livestock sector.[1]

The ESV asks an analogous question regarding research and management and the acquisition of technology, whether the enterprise has linked to IES in order obtain research, development and technological innovation services, yielding a slightly higher amount of 10.11%. This same survey asks the company if it carries out constant scientific work linked to its production process. The positive response came from 41.57% of the respondents. The obtaining of technology for productive processes in these companies can be seen in Figure 8.4.

As can be seen, the acquisition of technology is basically done through purchase. If we interpolate at the level of registration both the local purchase of technology (of foreign origin) and foreign purchase of technology (of foreign origin), the true value rises to 67.42%.[2] Table 8.3 reveals what happens if the latter value is broken down by the sectors of greater weight in the survey. Professional, scientific and technical services sector dominates in the purchase

---

1   The sum of the disaggregated values from both tables does not correspond to the global values for research and management (7.39%) and acquisition of technology (6.22%) on account of the fact that in reaching a global total at the level of each company's register, the appearance of various rubrics should only be counted once.

2   The sum of values is realized at the level of each company file, where the appearance of both rubrics is counted only once.

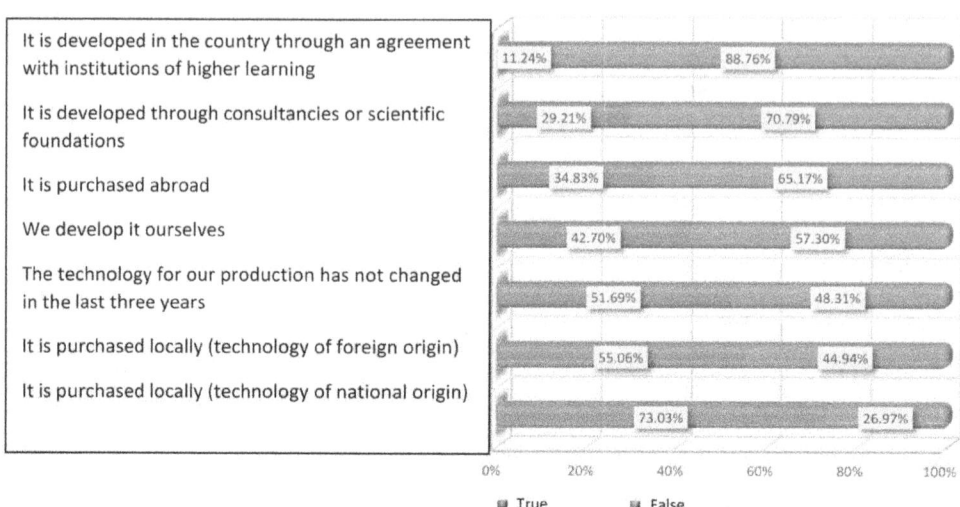

FIGURE 8.4 *Where has your company acquired the technology (licences for products or processes, machinery and equipment) necessary for your production process during 2011 and 2012? (nonexclusive responses)*
SOURCE: ESV

of technology effected abroad with 90.91%. Table 8.4 shows that this same sector is also the most active in terms of linking up to IHE at 27.27%. Finally, it is also the sector that exhibits the least technological obsolescence 72.73% (see Table 8.5)

Professional, scientific and technical services have the characteristic of focusing "mainly on business and have an economic impact on them. Most are specialized activities that were traditionally carried out internally by the same businesses and that are now purchased by them as just one more input. Its economic importance has been growing" (INEGI 2013: 18). As in the case of the manufacturing industry, these activities have the potential to increase links with IHE for furthering research and development (R&D) or the direct participation of IHE in this sector.

The ESV exemplifies the case of the *Soluciones para el Control de Recursos, S.A. of C.V.* that has developed a system for monitoring greenhouse gases with the Centre for Research and Advanced Studies of the National Polytechnic Institute (CINVESTAV-IPN). This system entails performing research and design of nanotechnology for the development of materials for injection of the plastic bodies of water meters, and the design of a pilot plant of manufacture and assembly with a lean manufacturing system of meters developed by the Autonomous University of Nuevo León (UANL) and the Corporación Mexicana de Investigación en Materiales SA of C.V. (COMIMSA).

TABLE 8.3 *Technology purchased abroad (licensed products or processes, machinery and equipment) needed for productive process during 2011 and 2012 by sector*

Sector	True	False
Construction	64.10%	35.90%
Industrial manufacturing	65.38%	34.62%
Professional, scientific and technical services	90.91%	9.09%

SOURCE: ESV

TABLE 8.4 *Technology developed domestically under contract with institutions of higher education (licensing for products or processes, machinery and equipment) needed for the production process during 2011 and 2012 by sector*

Sector	True	False
Construction	5.13%	94.87%
Industrial manufacturing	15.38%	84.62%
Professional, scientific and technical services	27.27%	72.73%

SOURCE: ESV

TABLE 8.5 *Technology for productive process has not changed over the last three years*

Sector	True	False
Construction	61.54%	38.46%
Industrial manufacturing	50.00%	50.00%
Professional, scientific and technical services	27.27%	72.73%

SOURCE: ESV

According to ENAVES, the main reasons why companies have not sought linkage with IHE in the most general sense boils down to administrative issues on the part of the latter, i.e., the supply of services potentially being offered by IHE is unknown as are the procedures for accessing them (SEP-CIDE 2010). The ESV supports this idea, adding that the perception of the respondents is that access to agreements with IHE can pose the risk of bureaucratic entanglement.

In the ESV, 62% of entrepreneurs indicated that they consider the Mexican state as having no interest in the development of its own technology for use in the productive process of Mexico's companies.

Conclusion

The data that has been presented reveals that Mexican entrepreneurs prefer to obtain the necessary technology for their production processes by purchasing what is largely of foreign origin. They also show that collaboration between IHE and Mexican companies in the development of technology remains incipient. Therefore, the trilateral networks and hybrid forms of organizational collaboration envisioned by the Triple Helix Model are almost non-existent. As long as these indicators remain unaltered, Mexico will remain an industrial colony of others.

It can be affirmed that there is a great indifference among the constituent branches of the Triple Helix, and in particular, business entrepreneurs do not consider higher education institutions to be a serious option for producing research and development according to their needs. These entrepreneurs need to stop conceding "their existence and enrichment at the margin of the advanced countries, or better put, at the margin of the development of these countries" (Figueroa 1973: 205) as the first condition for laying the material bases that can support building linkage.

CHAPTER 9

Proliferation of the Corporate Agro-Industrial Model in Latin America

Irma Lorena Acosta Reveles

Introduction

This study is derived from a group research project[1] that explores the social implications of scientific work as it is linked to the larger processes of production and the exercise of democracy. In this context, the present work seeks to contribute to this general theme from the vantage point of observing agricultural production in the region of Latin America. This will empower us to learn more about the effects, the larger potential, and the socio-economic, political and environmental challenges that science and technology represent as applied in the agricultural sector.

A broad overview will reflect upon the widespread and heterogeneous penetration of the Corporate Agro-industrial Model (CAM) in all of its complexity and shortfalls throughout Latin America. It quickly becomes apparent that the technological paradigm that sustains this model has failed to adequately respond to the problems of social inequality, exclusion, and poverty that typify the region. Our examination can help identify the model's fissures, structural flaws, limits and excesses, as well as the possibilities it opens for organized social political action at the supranational level. It is a veritable collation of tensions that occurs, both at the outset as well as throughout its development, in which conflicts eventually result that go well beyond rural areas.

Both capitalist and peasant economies co-exist across the region of Latin America. Our focus on this occasion will be on the capitalist pole, since it is there where the use and abuse of technological processes in the exploitation of land resources now prevail. While we examine this capitalist pole of production, which is the fastest growing in microeconomic terms and in macroeconomic resonance, we will of course be unable to omit frequent references to

---

1 The title of the larger research Project is "Science for Development and Democracy," funded by the Mexican National Council of Science and Technology, Basic Science Fund (SEP-CONACYT). N. 0105181.

the social and natural environments that it cohabits. Our focus on the operations of the foodstuff and agro-industrial corporations is nevertheless justified because they have come to depend upon this scheme of productive organization, supported by the governments of the region, as the most promising avenue to deliver economic prosperity. It is also a central focus because the value and global market prices of agricultural goods are set as a function of their operations.

This study is organized into three sections. First, we provide some brief historical notes about the social changes linked to the introduction of new agricultural technologies in Latin America during the Twentieth Century. Then, we offer a working theoretical frame that helps conceptualize the defining elements of the CAM. Finally, we set out to analyse in two parts the conflicts being generated by highly profitable but rather absurd modes of production.

Agrarian Technology and Social Change

What we are calling the Corporate Agro-industrial Model (CAM), which encompasses all of agribusiness, of the agricultural exports apparatus and of the whole paradigm of industrialized agriculture, has its roots in the wave of capitalist expansion experienced at the end of the 19th and beginning of the 20th centuries during the rise of modern imperialism (Acosta-Reveles 2006). Thanks to the developing productive forces, the way was opened to a second industrial revolution based on the energy potentialities provided by fossil fuels. Machinery came into the field by 1892, and in the 1930s fertilizers were being mass deployed. The scientific-technological revolution that followed relied increasingly upon non-primary production inputs being applied in the field. Seen in perspective, the last century was more significant in terms of changes to the production processes of agricultural goods than the rest of human history combined.

Progress, however, was neither linear nor free of setbacks. Along the way, applied biochemistry (initially through military technology that after World War II became channelled into agricultural technology) brought with it extraordinarily high yields and unprecedented optimism about the possibility of solving the problem of world hunger. The "green revolution" was built upon three pillars that lasted for nearly two decades: machinery, non-organic inputs and hybrid seeds. Soon afterwards, the environmental consequences and financial costs began to undercut faith in that paradigm, coinciding with crises in state developmentalism and rising oil prices, thus leading to a search for alternatives and shifting foci based upon newer scientific developments.

Nevertheless, applied biochemical science, industrial agricultural inputs and fossil fuels all remained at the core of agricultural techniques. Emphases shifted from seed selection to genetic manipulation, from tractors and basic harvesters to precision combines, and from raw materials of generic industrial design to increasingly sophisticated technological packages that were more specific in objectives. This was largely made possible thanks to advances in molecular biotechnology and information technology.[2] The use of artificial climates, specialized laboratories, satellite communications, enhanced marketing techniques, traceability, state of the art computer applications, all in the service of agro-alimentary globalization, now positioned the capitalist entrepreneur as the central agent in the process. This heralded the era of neoliberal policies.

By the early years of the 21st Century, most countries in Latin America were pursuing national strategies that were broadly open to foreign investment and agricultural trade. These countries were somehow convinced that this liberalized path of growth offered their productive bases the optimal use of the best lands and water resources. In practice, the penetration of the CAM occurs at varying velocities (on account of political and institutional reasons as well as its investment capacity and the potential for alternative uses of its territory). But invariably, in each country they can easily identify the presence of poles in the use of advanced agricultural technology that is specialized in either tropical crops, oilseeds, fruits and/or vegetables (Acosta-Reveles 2010). In the same policy scenario, but with a different organizational logic, we find the peasant type of farming which either by conviction or necessity likewise incorporates certain elements of the model in its operation. In fact, there are different dimensions of agribusiness and various administrative levels where the state itself has set the guidelines for connecting farming households to the cycle of capital (i.e., vertical integration), including their supplying essential labour resources and assigning functional formats to their objectives (e.g., contract farming, leasing of peasant lands, and service contractors).

---

2   García Olmedo explains that the definition of biotechnology covers all technologies mediated by a living organism or parts of it, be they cells or isolated enzymes. Under this definition is included the process of agriculture itself, as invented ten millennia ago, practices such as brewing in ancient Mesopotamia, and technological discoveries such as the latest forms of human insulin. It is therefore not appropriate to use the term narrowly to refer exclusively to the very latest advances in molecular biology. For the latter, it is more appropriate to use the term molecular biotechnology: one that involves handling of cells and organisms through their genetic material, i.e., that of DNA in the test tube (García Olmedo 2009: 124).

Components of the CAM

Outlining the fundamental elements of CAM production as operated by corporations helps to expose some of the reasons why it became installed as the dominant system of production, such that it attains the ability to dictate the rules of the game to the rest of the participants:

1. With regard to the internal organization of work, the CAM operates based on wage relations (capital-labour) in that it purchases and utilizes purchased labour time (not necessarily legally, e.g., adults, etc.) from which it obtains surplus value.

2. The CAM operates on private property rights and private usufruct of productive resources under rights that states and international norms protect. Corporate ownership of the means of production includes goods and services, tangible and intangible assets (land, machinery, supplies, technology packages, seed patents, and expert advice). Operation of the model does not require the private ownership of land (e.g., leasing, possession) but enjoyment of the resulting benefits in the form of profits is private.

3. The use of the land and labour may be extensive or not, depending on the type of crop, but it tends to be intensive in the use of scientific and technological resources, i.e., it presupposes the organization and exploitation of general labour, not only immediate labour. This requires important investments in order to start-up. However investment in this resource is not always exclusively private in that it can rely upon publicly sponsored infrastructure, research and development among its aspects.

4. The six basic practices that constitute the vertebral column of this dominant system of agriculture are: intensive farming, monocrop cultivation, irrigation, application of inorganic or synthetic fertilizers, chemical pest control and genetic manipulation of crops (Gliessman 2002: 3).

5. The agricultural activity in its various phases of production is strictly planned and evaluated in terms of risks. The aim is to get further and further in control of the production process, overcoming climatic barriers (plasticulture and artificial climates), shortening the period of ripening, altering the natural stages of the plant being harvested, protecting crops from pathogens and foreign agents, speeding its growth, intensifying selected aspects of the crop and lengthening its post-harvest lifecycle. These were aspects largely not pursued during the earlier era of the "Green Revolution."

6. Decisions and the fate of participants' products are guided by market criteria, especially global markets. Their stock levels are such that they have

a determinate impact on global prices, and their margins of utility allow them to place their products around the world while continuing to invest in scientific developments in favour of their competitive advantages. All of this is conducive to low production costs that reinforce their traditional comparative advantages of cheap labour and land access, leading to extraordinary profits that are further elevated when they coincide with exceptional soil quality.

7. In the continual search for better investment opportunities, they resort to a pattern that resembles industrial relocation, i.e., a geographical mobility of agro-capital with different agricultural products based in flexible specialization, allowing movement across varied links of agro-industry and available inputs, the deployment of machinery, soft technologies, and movement across exceptionally cheap and docile sources of labour that are pushed and pulled into availability.

8. Linkages are established with other economic branches (supply inputs, agricultural machinery, transportation, refrigeration services, infrastructural networks, marketing, etc.) and their experiences in the field, providing feedback to research centres, the seeding, fertilizer and pesticide industries, and so on. This ultimately reaches into participation in financial markets, such as commodity futures that in turn fuels commodity speculation.

Such is the nature of the hegemonic model. The conditions for any optimal operation are supported by public policies that reflect their interests: operators receive privileged budgetary support, institutional considerations for investment, and facilities in the use of land and local labor. Their aims are projected as if they represent the general interest via the promise of trickle-down economic benefits by capitalizing on local land resources, creating jobs and contributing to a nation's gross earnings.

With the CAM, peasant agriculture becomes ever more subordinated to agro-industry even while it adds value to its products. The paradox of the model is that agriculture is no longer the primary link in the chain of production, but instead just one more link in a complex system that articulates all of its branches right up to the consumer, that centre of gravity around which orbit the seed industry, agro-chemical production, agricultural biotechnologies, the processing sector, agro research centres, packaging industries, and supermarket chains.

The co-existing model of peasant agriculture, no longer in harmony with agribusiness, continues to operate under a different internal order of production, namely, a family-based household economy that conserves simple, non-extended reproduction as the lynchpin of its decision-making. This system generally operates on a small scale with a precarious material and technological foundation.

Dwarfed in its overall economic impact, peasant agriculture achieves a vertical integration in the CAM where family farmers fit into the scheme as labourers, consumers and to some extent as suppliers provided they can manage to comply with the guidelines and requirements established by agro-business. Indeed, we can observe how in recent years, international agencies such as the FAO have increasingly incorporated initiatives into their support strategies aimed at linking family-based agricultural producers to global foodstuff chains:

> Foodstuff processing industries add value while increasing the demand for agricultural products, thus contributing to the reduction of poverty and improving the food security throughout rural areas. They offer employment opportunities in off-farm activities, such as in the handling, processing, packaging, storage, transport, and marketing of food and non-food agricultural products.
>
> The FAO works in consultation and collaboration with its state members to address their specific needs related to the development of foodstuff industries by sponsoring training and technical support. Our capacity-building activities are carried out through field projects and training programs. Informational materials and training activities comprise a wide variety of topics and are distributed by electronic or print media and target diverse audiences, such as public and private sector organizations, universities and technical institutes, NGOs, researchers, instructors and various other participants in the chain of post-production. The main beneficiaries of direct technical assistance that we provide in the field include micro, small and medium-sized businesses which are the main processors of food and agricultural products in developing countries.
>
> DA SILVA et al. 2013

Cracks and Tensions

Regarding the virtues of the CAM as described, both regional data as well as figures from various countries have constantly alluded to the growth in the volume and commercial yields, the expansion of exports, and the development of new food products from the region that gain prominence due to their unique qualities or prices arising from specialization. There is no better evidence of this success than the magnitude of capital that transnationals have amassed under this paradigm of production, including companies such as Monsanto, DuPont, Syngenta, Nestle, Pepsico, Coca Cola, and others. While these large transnationals have grown throughout the developed world there has been a similar development of large national firms, in virtually every Latin American country, that have arisen on the increasing volumes of exports or by virtue of their positioning products within key sectors of the gross domestic product. Examples of this include Expofrut in Argentina, Subsole in Chile, Gruma (Maseca Group) in Mexico, and Sadia in Brazil.[3]

As a whole, Latin America has achieved great success in dramatically advancing the use of its land for the production of grains and oilseeds under CAM parameters, using high-tech means of elevating productivity and export, and even contributing in some cases (especially Brazil) to innovation in this sector. Macroeconomic data demonstrates that the region continues to be a large exporter of primary commodities, and that its agricultural success is being spearheaded by several countries (e.g., Brazil, Argentina, and Colombia). The problem, in short, is that the economic benefits have remained highly concentrated, while in the social and ecological spheres the outcomes have not been positive. This system of development has not contributed to the reduction of poverty, national food security, containment of migration, generation of employment, or environmental sustainability. On the contrary, this regime of production has been responsible for territorial conflicts and violent confrontations over the control of natural resources, disruption of ecosystems, massive human displacement, and the widespread destruction of small producers. Nor has it helped to alleviate disease or extend life expectancies. In order to render these issues more manageable, we shall examine them more fully in two separate parts, namely, as social and environmental problems. We are mindful at the outset, however, that this separation is contrived given their intimate and permanent interconnections.

Agriculture and the Natural Environment

In principle, agriculture is a social activity. Shaped by culture, it is a conscious intervention of nature that seeks to domesticate it. But today as never before, it is subjected to what from the dominant view are seen as *human* necessities. In this moment of history, these needs are identified as the needs for development of the capitalist system. Only by manipulating nature through available technological means does it seem possible to respond to the challenges of an

---

3   Agro-foodstuffs companies present in Mexico include: Monsanto, Cargill, Archer Daniels Midland, Tyson Foods, Dow Chemical Company—and its subsidiary Dow AgroSciences, Bunge, JBS S.A.—and its subsidiary JBS US.

unprecedented population growth in an increasingly urbanized and interconnected world. This amounts to adequately feeding a world population that between 1960 and 2010 grew from three to seven billion people and which is expected to reach ten billion by 2045.

The possibilities to continue increasing crop production from the land have historically developed along three tracks: (1) by expanding the area under cultivation; (2) by improving the use of cultivated land, e.g., by reducing the rest periods for soils, double cropping, etc.; or (3) by multiplying the crop yields per unit area of land under cultivation. In this scenario, the application of scientific technology seems the most promising route as it has proposed to surpass the limits otherwise imposed by nature. But does this scenario really open the way to the future? The domestication of nature has so far registered social advantages and profits, but at the same time, it has done so with certain costs. The attempt to subdue nature, adulterate it, and squeeze it past the limits is to effectively go against it and drag it headlong down a dead end route. After decades of technological advances, huge investments, and widespread geographical distribution of genetically engineered crop varieties, it is manifestly clear that modern agriculture has not delivered on the promise of feeding the planet's inhabitants. And what is equally well known is that the problem is not one of a technical nature or the lack of capacity, but rather one of politics, of decision-making and of vested interests.

If we observe how the CAM model outlined above works in practice, it can be seen that farming is not a trade or a means to sustain life, but rather one more link in the instrumental goals of the transnationals. Agro-industrial corporations that sponsor ongoing scientific development have subjected nature to their own ends thus diverting a once spontaneous vocation. They have been abusing soils with poisonous chemicals obtained from non-renewable resources while at the same time enabling the creation of weeds and microorganisms that are increasingly difficult to eradicate, disrupting the climate, depleting water reserves, and exterminating diverse forms of life through monocrop cultivation.

This is the case of forests that are being replaced by soy, rice, corn, and grain cultivation, all of which is of dubious quality for consumption, but which nevertheless is destined for consumption for cattle, foodstuffs, and even to produce energy via biofuels. The demand for these commodities continues to grow for which the natural environment must continue to produce ever more. Furthermore, industrial agriculture that is sustained through an increasing consumption of fossil fuels is expected to be capable of generating clean, inexpensive and large-scale energy sources that can come to replace precisely these fossil fuels. Such an energy challenge in this century could not be more absurd.

Another aspect barely spoken about is that large scaled agriculture and livestock industries are among the largest emitters of greenhouse gases such as carbon dioxide, nitrous oxide and methane (FUHEM Ecosocial 2013: 4). Water resources are likewise at the service of business and in Latin America there are serious shortages, pollution and conflicts over this precious liquid resource despite the fact that the region possesses the largest reserves of clean drinking water in the world. Since everything is now valorised in market prices, water is being registered in national accounts as "virtual water" and the "water footprint" as part of wider *water markets*.

Virtual water is regarded as the amount of water imported and exported in agricultural products and services, including amounts contained in products and additional amounts consumed in their production. The water footprint is calculated by the amount of water consumption needed to maintain a human hydrated through the goods and services consumed throughout his or her life (Garrido 2013: 145). Both indicators have emerged out of the well-founded fear of scarcity of this strategic liquid in the face of wasteful production practices by the CAM.

The two axes of greatest influence on global water resources revolve precisely around agricultural trade and global climate change, with the latter significantly affected by the former as a result of changes in the productive use of soils. With increased population and changing diets, more water is required to produce the additionally needed food and this can be expected, especially in some areas, to result in social conflicts. In the case of Latin America and the Caribbean, expert estimates suggest the possibility of further increases in the area under irrigation from 19 to 78 million hectares by 2030. As can be seen in Figure 9.1, such resources are available. What is not clear is whether there is sufficient land suitable for irrigation, sufficient capital available to transform those lands, or the possibility to avoid the kind of conflicts that may result in these rural areas given the existing populations and producers who presently use these resources.

Two additional practices typical of industrial agriculture operate at different rhythms than those of natural life cycles. Namely, "smart agriculture / high-velocity agriculture" and the production of grains, oilseeds and other products derived from genetically modified organisms, i.e., the GMOs that we have all heard so much about. These practices are most commonly employed in the cultivation of pumpkin, alfalfa, beets, cotton, soybeans, corn, tomato, papaya, poplar, canola, potatoes, and sweet peppers.

In high velocity agriculture, cultivation takes place under a regime of strict control and planning where practically none of the processes involved are left to chance. This is the case with hydroponic tomatoes and green leafy

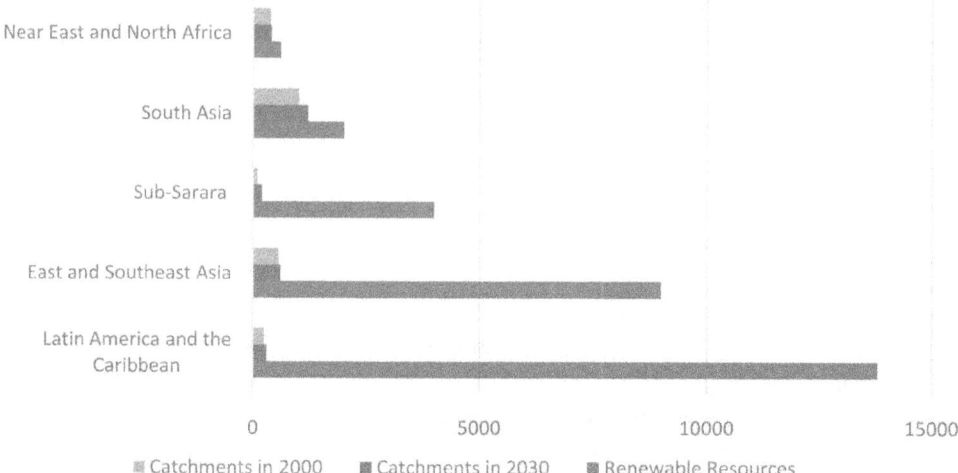

FIGURE 9.1 *Use of water and renewable resources in various regions of the world (in cubic Km)*
SOURCE: GARRIDO (2009) WITH DATA FROM THE COMPREHENSIVE
ASSESSMENT OF WATER MANAGEMENT IN AGRICULTURE (CAWMA) (2007)

vegetables fed essentially with nitrates. Here we may speak of virtual factories of vegetable production.

GMOs have meanwhile invaded the Americas extensively. The United States has taken the lead with more than 73.1 million hectares planted with transgenic crops in 2014, followed at a considerable distance by second place Brazil with 42.2 million hectares under transgenic cultivation. These giants are followed by: Argentina (24.3 million), India and Canada. In the case of Brazil, the figures cited had sharply increased from only 11.9 million hectares over just a three-year period (AGROBIO 2015). Notably, the exceptionally fertile lands of the Amazon Basin and the plains of the Pampas regions have been facing the threat of erosion of their resources over the course of three decades, per expert estimates (see Table 9.1).

After Argentina and Brazil, the Latin American countries that follow suit in extensive GMO use are Paraguay, Uruguay, Bolivia, Colombia, Chile and Honduras. Given the social and political resistance to the proliferation of transgenics, precisely because it threatens and contaminates native crop varieties, it is supposedly the case in Mexico that genetically modified corn varieties are being grown only for experimental purposes. Transgenic soybeans and cotton, however, are being openly cultivated. According to the Association of Agricultural Plant Biotechnology (AGROBIO), Mexico ranks sixteenth among global producers in the practice of GMO planting, or in sixth place in Latin America, having placed 0.2 million hectares under cultivation by 2014. AGROBIO,

TABLE 9.1   *Data on global transgenic cultivation (2014)*

Principal producers		Products in the region	Latin American country / products
1. United States	11. Bolivia		Brazil: corn, soy and cotton.
2. Brazil	12. Australia		Argentina: corn, soy and cotton
3. Argentina	13. Philippines	Soy	Paraguay: soy
4. India	14. Myanmar	Corn	Uruguay: soy and corn
5. Canada	15. Burkina Faso	Cotton	Bolivia: soy
6. China	16. Mexico	Canola	Mexico: soy and cotton
7. Paraguay	17. Spain		Colombia: cotton
8. Pakistan	18. Colombia		Chile: corn, soy and canola
9. South Africa	19. Chile		Honduras: corn
10. Uruguay	20. Honduras		

SOURCE: ASOCIACIÓN DE BIOTECNOLOGÍA VEGETAL AGRÍCOLA

itself a leading developer of this type of GMOs, describes itself as a non-profit association dedicated to inform, educate, disseminate and promote modern agricultural biotechnology, and works closely with organizations interested in research, development, production and marketing of these resources:

> We believe in the rights of every citizen to be informed, to access the bene-
> fits of biotechnology and to decide upon their acceptance. For this reason,
> we provide accurate, timely and scientifically substantiated information
> in an ethical manner.
>
> AGROBIO 2015

The deforestation that accompanies the concerted search for opening new fields for growing grains and oilseeds is effectively another factor in the loss of biodiversity and the generation of climate imbalance and water shortages. Deprived of any possibility of participating in agro-industry given its costs, the regional peasant economy incorporates some of its practices (seeds, machinery, biochemicals) that in turn further erodes and adversely affects their natural resources. The accumulated knowledge and human intellectual work done through science and materialized in these sophisticated means of production have established a pattern wherein the goods that nature provides are no longer the sole material and tangible starting point of the agricultural production process. Instead, the very same products that have been processed, adapted and transformed are now those being sought after and appropriated for buying and selling.

In short, the environmental issues regarding the manipulation of nature that arise within the framework of the application of the CAM has yielded the contamination of water, the depletion of groundwater, widespread deforestation, and overspent and eroded soils that have been domesticated to serve purposes other than those dictated by their natural vocation. This has created a world of pest resistant plants, genetically mutated microorganisms, a damaged and deteriorated biodiversity, and a depletion of non-renewable resources.

Agrarian Economy and Social Deficit

Just as in the realm of nature, the spread of CAM has been a source of tensions in different fields of social life. In some cases, it has resulted in simmering conflicts that gradually build to a head, provoking organized responses that have resonated and echoed in distant places. We will proceed to discuss some of the most significant cases.

First at the macro level, we can see the disequilibria in national accounts that have resulted from the need to systematically import primary agricultural goods. Indeed, one might think that with the CAM and all of its technological potential, the problem of ensuring adequate foodstuff supplies would be completely overcome. But this is not the case. The regional agricultural trade balance remains in deficit as agricultural trade surpluses can be found only in a few countries. On the other hand, even if one assumes an adequate, timely and accessible supply of foodstuffs to be in place for the population of a given country, albeit thanks to imports or the substantial increase of crop yields, the matter is not resolved with the simple presence of products in the market. This leads to the classic question of social redistribution of national income.

Secondly, we can see at the micro level that with the rise of agribusiness comes the loss of profitability of many other productive units and their expulsion from the business cycles in which they once participated. It is well known that the farms across the region are quite heterogeneous in size and type. Hence, the price dynamics set into motion by the CAM, based on its technological and organizational practices, places the less efficient producers in crisis, first by reducing profits and ultimately with their resignation from farming as a livelihood. Expanding the amount of land under cultivation, improving the quality of inputs, gaining access to better services and innovations, or other measures aimed at reducing costs are simply unavailable as options for many producers.

Third, the prevailing technological schemes in place today have placed small farming businesses in a precipitous decline on a global level (Garcia 2012: 401) and Latin America has proven to be no exception. Nor has the quality of agricultural work improved under the CAM. On the contrary, the precariousness of agricultural labourers has become accentuated (Acosta-Reveles 2010). It is an activity that continues to be marked by high levels of risk of illness and physical injury of workers who chronically suffer from a lack of unionization, high instability of work, a need for frequent relocation, illegal status, and overt forms of repression.

Fourth, we must consider the social conflicts that arise from land expulsions and the hoarding of the resources of production such as land, seeds, water, forests, and the native genetic materials of plant or animal origin. With regard to land resources, neoliberalism and its support of CAM has resulted globally in the trend towards a re-concentration of rural areas, especially of the best quality lands, those accompanied by the best infrastructure and/or those found in the most propitious locations. Cultivation of areas that enjoy access to water, minerals and biodiversity, are being steadily absorbed, although not necessarily in ownership, for the benefit of private farming that is specialized in export items. This farming is extensive, when merited by the type of crop and intensive in its use of technology. In this process, the most vulnerable sectors of family farmers retreat or are expelled from the areas deemed most desirable for new uses, including for speculative purposes.

With the global rise in food prices of the 2007–2008 period, the issue of land re-concentration increasingly took centre stage, drawing attention to the purchase of vast parcels in developing areas by Middle Eastern or Far Eastern countries. These countries included Saudi Arabia, China, India and South Korea, perhaps due to a climate of fear about an eventual shortage. However, the context in which these inclinations for land grabbing should be read is one of geopolitical confrontation where the accumulation and control of strategic resources is key. This is coupled with the fact that a position of technological dominance for private interests allows for production with significant advantages on the world market. A synthesis of various critical studies on the subject suggests that rather than focusing on the concentration of ownership per se or on the scale of the phenomenon, the political component of domination stands out at the root of the process and its impact on the economic, social, environmental and cultural dimensions:

Land grabbing therefore epitomizes the change that is increasingly occurring and what it signifies for the use of land and associated resources (such as water) by family farmers who produce on a small-scale and in a labour-intensive manner for their own domestic consumption and for local markets. The land is being appropriated for use by large scale, capital intensive and resource draining agriculture, such as monocrop industries, the extraction of natural resources and in service of major infrastructures for the generation of electricity, the likes of which are integrated into ever expanding infrastructures that link the zones of extraction to metropolitan and foreign markets.

FUHEM Ecosocial 2013: 2–3

Fifth, it can be seen in strictly social terms that the spread of the CAM and its high-cost technological pattern tends to deepen the productivity and income gap within agriculture. The fact that the model becomes imitated in some of its aspects such as the use of certain inputs, techniques or organizational schemes does not imply that the benefits are becoming shared or that progress is socialized. In the current socio-economic order where the welfare state has been reduced to a minimum expression, the effective exercise of social rights in the countryside or in the city increasingly becomes reduced to the purchasing power of consumers. In turn, the level of individual consumption, of family consumption, or of the small scale units of production implies a status of economic inclusion, be it either by work in the form of wages or otherwise, or by incorporation in regular economic cycles, i.e., when a domestic or capitalist enterprise operates and becomes profitable. In the region, however, the imprint of the neoliberal era in agriculture is one of increasing exclusion and of reproduction in ever more precarious conditions that ultimately threaten basic subsistence. A virtual mosaic of various historical deficits, there are two critical issues that particularly emerge out of the deployment of the CAM, namely, basic nutrition and sanitary levels.

Regarding the issue of food and basic nutrition, the debate as well as the problem itself has gone through several transitions. From the beginning the institutional posture put forth (by the FAO) referred to the priority to protect food security in the face of insufficient supplies. Then the institutional posture focused on social and nationalistic reaffirmations of food sovereignty (around 1996) that included issues beyond hunger, mapping adequate supplies as a problem and referencing social variables such as nutrition and infant mortality rates. The positioning of the issue on the political agenda of the 21st Century likewise reminds us of the food crisis situation that has been anticipated since 2005.[4]

---

4 Up through 2007, the concept of food security continued to broaden and become further

By that time, the penetration of agribusiness and the CAM had extended throughout all countries of the region, altering conventional agro-production structures, including those producing foodstuffs and especially basic grains production and transforming them into a specialization oriented towards foreign markets. Severe changes in the use of land, the incorporation of industrial inputs, mechanization and dismantlement of sustainable practices had all taken place. As these changes converged with the changes of the Western diet towards meats, cereals, and industrialized food products, particularly in the higher-income countries, the fall in international grains reserves and the increased demand for biofuels all conspired to produce instability in the global grains trade. The speculation on cereals in futures markets based on the abundance of non-productive capital, put nations and producers alike on alert. In the first decade of the 21st Century, a historical counter trend developed where food prices experienced increases, with spikes in 2007–2008 and 2010–2011.

The uncertainty that at the level of governments and businesses led to land grabbing and territorial control has at the social level led to greater collective action being organized to protect food rights. This has produced struggles to recover organic production, to preserve biodiversity, to improve local markets and facilitate access to productive resources, to move away from the use of GMOs, and to abandon food production for biofuel purposes. These movements seek an inclusive and sustainable agriculture that can likewise alleviate the problem of climatic change.

The other critical focal point of conflict that will just be touched upon here revolves around the issue of sanitation. The demands for increased production and productivity that should lead towards a successful model of farming today have miscalculated and underestimated the importance of creating safe products. Likewise ignored have been the rights of rural dwellers to have a clean environment, free of pesticides and other agro-contaminants, and the rights of labour to be free of the ongoing risk of illness and injury in the fields. Industrial agriculture, with its technological packages, has been told repeatedly that fumigation practices are poisoning rural people, polluting their rivers and lakes, and causing illnesses. So it is that amidst the abundant bonanza of highly productive territories there exists widespread discomfort and bitterness. Also proliferating are organized actions of protest and advocacy for health, denunciations

---

refined. The more recent proposal in the Nyéléni Forum held in Mali presented a notion that goes well beyond accessibility or availability in order to incorporate issues of rights and decision-making regarding the food system itself.

of the pollution of water, soils and air, claims of damage to the environment and health, and protest against occupational diseases and the harmful effects of aerial spraying.

Millions of farmers and their supporters all across the region have declared legal and political war on corporations such as Monsanto and in some cases judgments have been won, such as in France, Argentina, Brazil, Mexico, Chile, Canada, and Ecuador.[5] Most of those who dare to take on the transnational giants find themselves mired in a fight where the legal, political and economic system has been designed to protect corporate rights. While there have been some cases won, they are relatively few in the face of the large numbers of legal cases actually filed. Large agro-companies can cope with these actions given the magnitude of their profits. It has become standard practice for these transnational corporations to budget in the cost of litigation, indemnification, legal fees, insurance and so on.

The types of companies like Monsanto, DuPont, Syngenta, etc., have also been charged on account of their continued claiming of privileges on the use of seeds. They seek to establish patents over all new varieties and other previously unregistered native varieties. One of these transnationals puts it best on their corporate website:

> Monsanto invests more than US$ 2.6 million a day in research and development, something that ultimately benefits farmers and consumers. Without the protection of patents, this would not be possible (...) no business can survive without being paid for the products it generates.[6]

As for the consumer, the debates about the safety of genetically modified organisms (GMOs) are seemingly endless, including about the need to accurately label their use in products. The scientific doubts that exist are effectively used to protect producers. Nevertheless, some countries have said no to the market-

---

5   These protests continue to intensify. In 2012, a movement of 300,000 US farmers launched a suit against Monsanto. More recently, five million Brazilian soy farmers were litigating against the same transnational corporation, not just against the health problems being caused by the pesticides being used by the Company, but also to decry the impact that its practices were having on farmers who have lost or are losing their lands on account of debts they hold to Monsanto for royalties that they have been obligated to pay for the use of their patented seeds, even though they had not purchased them directly but had instead garnered them from harvested plants grown with the seeds.

6   See the website of Monsanto: www.monsanto.com/global/es/noticias-y-opiniones/pages/porque-monsanto-demanda-a-agricultores-que-reutilizan-las-semillas.aspx.

ing of GMO-produced goods (such as Spain and Germany). French and Austrian research has documented negative effects of GMOs on human health, not only by the variety of adulterated seed, but also by the use of bio-chemicals and growth-inducing hormones.

Conclusion

There are conflicts brewing in the regional landscape that are far from being insignificant. The substance of these conflicts reflects points of contention that are inherent in the CAM and which are not amenable to technical solutions. There is obviously substantial rancour in the population involving the disputed use of patented seeds or resources, regarding expulsion from their lands or over the threats to native patrimony. There is also protest over the effects of the CAM on human health, as well as on the part of those who are losing their livelihoods.

The legal actions taken against agro transnationals have from time to time resulted in victories. Thousands of farmers in Brazil, Chile, Mexico, and Argentina have taken up this cause. Some governments have also promoted developments that are being challenged by their citizens and this offers greater latitude for democratic actions against the use of such technologies (e.g., in Germany). Struggles over water resources, food sovereignty, movements demanding a halt to pesticide spraying, others protesting in defence of biodiversity and for the adoption of alternative (agro-ecological, biodynamic) practices and the preservation of family farming are all observable in the region. There are instances where cases have taken decades to adjudicate and charges have been openly made against a model seen as predatory and genocidal.

International norms often reflect established interests such as the Cartagena Protocol which calls for flexibility in the sense of "the technology is there, so whoever wants to, can use it." In any case, Latin American agriculture is a minefield with the growing discontent that is present, and a time bomb is being produced by the hyper exploitation of nature. If the potential opposition does not emerge from the organizations of those most affected (which are considerable in number), there also exists the possibility of change on the horizon via the loss of profitability due to the depletion of energy sources. The energy matrix that sustains this model rests upon fossil fuels.

There are reports circulating that the existent technological developments of today may no longer be profitable in the context of future energy crises. The relevant fact is that the global deployment of agribusiness has generated scenarios of contestation, and problem areas for the profitability of global capital. Since the scope of the system enables it to transcend the nation state, the challenges of social movements and organized collective agency may also reach a point of a regional or global movement as suggested by the transnational social movement organization *Via Campesina*. Theorizing suggests that the overall scenario should return to that level. I believe that at this time there is a commitment to do so by critical social science, to lay bare the truths that are hidden and to render visible that which matters. Unfortunately, the science that has been specifically applied to agriculture has been commandeered and financed by the private sector and as such is little more than a rentier science.

CHAPTER 10

Well-Being and Happiness: Conditions for a New Conception of Development?

Ernesto Menchaca Arredondo and Leonel Álvarez Yáñez

Introduction

This study seeks to demonstrate the complexity involved with analysing well-being, beginning first with its own conceptualization and then a review of the different modes and ways that have been employed globally in order to try to measure it through the construction of various indices and social indicators. We will outline how subjective well-being has become an item of interest for the development of political science and to that end, we present some of the main evidence that has been found that shows its importance in diverse spheres of the lives of people.

There is little doubt that traditional indicators for measuring development and/or social progress remain insufficient for explaining the new paradoxes that exist between the subjective and objective factors of well-being. To bolster our argument, we propose a complex measurement scheme that allows for the concurrence of factorial analysis and relational systems that display an awareness that subjective well-being should be considered as a key element for a new conception of development and the progress of societies.

Forms of Well-Being

In the pursuit of well-being, we can etymologically extract its Latin roots *bene* and *stare* that move us to rethink the meaning of "well" and "being" as a tangled set of resignifications that it is necessary to dissect. On the one hand, "bene" refers to various related expressions such as goodness, abundance, and perfection, all referring to something positive and desirable that immediately presents us the dilemma of whether something is good in and of itself, or whether it is something that participates in a state of "well-ness."

On the other hand, the *stare* alludes to the word "state" which denotes a mode of being within a certain reality, to be in a certain situation or a certain quality of being.[1] It can be analysed in comparison with the idea of "being like that," or in other words, an essence, existence, entity, habit, condition, etc. *Stare* also alludes to the concept of "status" as it was used in the Middle Ages, related to expressions such as *status naturae* or "state of nature" in a more theological and anthropological sense. All of this suggests that we are facing a complex conceptualization of a term that in some sense we all seem to understand but one that leaves many questions for a deeper and more adequate analysis.

To study well-being, if we see it as a term, a notion, a concept, an expression or a property of something, it suggests multiple implications that spins the analysis once viewed from one to another perspective. At least two configurations emerge, namely, the objective and the logical. If well-being is understood as a mental phenomenon alluding to a subjective view, this normally take analysis towards the psychological end. If it is understood as a "formal object" that is distinct from the expression of the phenomenon, than it can be analysed from an "objective" or formal approach using a systematic logic. In our approach, we are inclined to take a comprehensive view that integrally alludes to both the subjective and the objective aspects.

If we make a literal analysis of the meaning of "well," it can be seen as an adverb of a superlative and comparative nature. At the same time, it can be read as an adverb of quantity that adduces a certain value, or alternatively may speak to an emphasis on a lack of value. Our first position can therefore be to recognize that "well" can take various forms, as something that is inside of us and something that is outside of us. Therefore it is something that is both immanent and transcendent, dialectically speaking, that breaks the barrier of subjectivity and objectivity. That which is within us we refer to as how we feel, what we think and what we desire. That which is outside of us we call nature (the other or the others), also dialectically, because we already know that we ourselves form part of it. These aspects are joined to the idea of "being" relative to a certain person, or group, community or given society, to a certain historical time and a way of seeing the specific life. Thus, we can now start to see a preliminary framework that may seem more like a small conducive thread regarding well-being that does not pretend to exhaust all of the problems it raises, nor try to highlight all of the difficulties that this conception offers.

---

1   The word "state" can be translated into the Greek category of laid down or seated. It can also be translated as "situation" or "posture." But it might be translated into the Greek χάσχειν as "cut." And it has also been interpreted to mean "passion." This possibility of the same term being used to signify various meanings indicates that there are various modes of "stare," i.e., different states of being.

Global Views on Well-Being and Its Indicators

The Organization for Economic Co-operation and Development (OECD) published a series of studies on the standard of living of the world's population, using what they call the OECD Better Life Index (OECD 2013). These studies recognize, on the one hand, the need for micro-level analysis. Macro-level analyses cannot accurately reflect what is actually happening in a given area of a particular country. On the other hand, they recognize the difficulty and genuine challenge of analysing the well-being of people as a multidimensional concept, especially when trying to obtain a reading of the entire set of indicators in a single value. However, the use of these types of comprehensive indices helps to provide an overview of the patterns of well-being that are occurring in different countries.

The factors that influence the quality of life of people and that provide human satisfaction are an important part of what we call social welfare. It is a condition that cannot be directly observed, but rather ascertained through various formulations and comparisons across space and time. Obviously, this concept has a very high subjective content because it includes an appreciation for the economic goals and aspirations of individuals. The concept of well-being and the practical ways of measuring it has been the subject of many debates among public policy makers and international organizations such as UNESCO, OECD and the UN.

Conventionally, well-being has been quantified by measuring the material goods and services produced by a country and dividing it by the number of its inhabitants in what yields the per capita income. That in a way was already surpassed by Amartya Sen's vision of social progress, understood as the systematic elimination of social deficits from which an index of social progress was conceived in terms of eliminating these shortcomings.[2] The discussion has focused mainly on the type of factors to be included or excluded in determining the standard of living of a population. For example, some indices include unemployment, marginality, poverty or a certain type of social dysfunction. On the income distribution side, there is discussion on how to measure its distribution amongst the population of a given country. At the same time, there is discussion about the factors that contribute to the increase of GDP per capita, such as productivity, employment rates, the number of hours worked, etc. Nevertheless, these measures only partially capture the totality of well-being.

---

2   See Desai, Sen and Boltvinik (1998) and Sen (1999).

Yet another index that attempts to measure the level of poverty in a country is the Human Poverty Index (HPI), developed by the United Nations for countries with developing economies. This index includes some of the following aspects: the probability of being born but not living to the age of forty; the rate of non-literate adults; the average rate of the population without stable access to a quality water source; and the average rate of underweight children.

Various other indicators that try to measure the well-being of a population have been constructed by diverse institutions. To list a few, there are the Index of Sustainable Economic Well-being (ISEW);[3] the General Progress Indicator (GPI); the Human Development Index (HDI) of the United Nations; and the Fordham Index of Social Health, the latter of which measures 16 different indicators including the rate of mortality, child abuse and poverty, suicide, drug use, school dropout rate, average earnings, unemployment, health coverage, poverty among the elderly, homicide, housing and social inequality. One further indicator, the Index of Economic Well-Being (IEWB), considers aspects such as family savings, household accumulation of tangible capital and the value of housing in addition to attempting to measure the sense of future security.

In addition to all of the attempts to incorporate indicators of well-being, there have also been attempts to measure the lack of it, as a different vision on how to attend to the problems of populations. These alternative approaches have given rise to a series of discussions and analysis about human development and poverty. In this vein are the following indices: the Multidimensional Poverty Index (MPI) that since 2010 has supplanted the Human Poverty Indexes (IPH); the Poverty Index or independent poverty indicators (HPI-1), developed since 1998; the Human Poverty Index for selected OECD countries (HPI-2); the Human Development Index regarding gender created in 1996, the Gender Empowerment Index (IPG); and the Index of Material Deprivation.[4]

The previous considerations have attempted to measure well-being and the feeling of satisfaction on the part of populations. This latter aspect, in particular, is also known as subjective well-being, and has inspired a series of methodological discussions regarding the development of techniques that

---

3   Based on the ideas initially developed in 1972 by Yale economists W. Nordhaus and James Tobin in *Measure of Economic Welfare* and further developed in 1989 by Herman Daly and John Cobb. The main objective was to replace the GDP with an alternative method of quantification that increased the well-being component. See Daly, Cobb and Cobb (1994).

4   Applied in Great Britain in 2010, this index included an indicator of poverty in terms of income and material deprivation in an attempt to improve the so-called Complementary Poverty Indicator developed in the US in 2011.

can better and more accurately measure both factors in a more coherent and homogeneous way. This is a problem that still remains to be resolved.

In recent years, a subjective well-being approach or focus has been prioritized in such studies of happiness as those by Kahneman (1999), Hills and Argyle (2002), and sociologists such as Veenhoven (1984). Among economists, research has been enhanced by what has been termed "the economy of happiness" (Rojas 2009). There are also two traditions in the epistemology of well-being which have been called imputation and presumption. In the case of imputation, it is very common to judge the well-being of people through third parties, often imposed or classified by the researcher or expert. In the case of presumption, the approach to well-being is through the way that people actually experience or manifest their enjoyment of their well-being.

The subjective well-being approach is based on surveying people directly about their well-being. In addition, questioners can inquire about happiness, life satisfaction or any concept related to the well-being of the person. The important thing is to be informed by people about the well-being that is the object of interest. Clearly, it can be observed that there are differences in the information that is obtained, depending on the objective being pursued, although many studies take these concepts as synonyms and they fail to distinguish between happiness, life satisfaction and well-being.

In general, we can point to three ways of approaching the study of happiness. A first approach consists in the direct study of the happiness of human beings. A second is the study of the relationship between happiness and economic variables such as income, unemployment and inflation. And a third style of approach consists of using happiness as a proxy for utility. Each of the three approaches have their own problems in applied analysis and require a diversity of resources and methods to successfully obtain verifiable identification.

Without doubt, the "economy of happiness" provides a methodology for assigning values to external events that can be considered as another approach within the traditional methods of quantification, as well as the contingent valuation approaches used by Ada Ferrer-i-Carbonell and Mariano Rojas.[5] Nevertheless, there remain significant gaps and doubts about the most adequate ways of analysing well-being or happiness.

---

5   This approach has also been used to rate illnesses, to calculate the income compensation necessary when a person changes residences, as well as to calculate the compensation necessary to add a new person to the household, in a way that can calculate scales of income equivalence. See Ferrer-i-Carbonell (2011).

The Subjective Well-Being of Mexicans

We now turn to take a general look at subjective well-being in Mexico. According to INEGI data on all people between the ages of 18 and 70 living in Mexico (INEGI 2012b), 47.3% indicated they are satisfied with their lives, 36.1% are moderately satisfied, 11.8% are little satisfied, and 4.8% are dissatisfied with their lives (INEGI 2012a).

On a scale of 0 to 10, the mean value of satisfaction with life given by the population group under study was 8.0. The aspects or spheres of life in which Mexicans show greatest satisfaction on the scale of 0 to 10 are: family life 8.6; autonomy 8.5; health 8.2; and affective life 8.2. In contrast, the worst qualified aspects are: economic situation 6.5; country 6.8; available spare time 6.8; and education 6.9. Similar to what has been reported for other countries, the relationship between life satisfaction and age is in the form of a "U" where the highest level is between the 18 to 29 years of age group (8.1) and the lowest (7.9) was indicated for the 45–59 years of age group. In between were both the 30 to 44 years of age group and the 60 to 70 years of age group with a value of 8.0.

Satisfaction with life was found to be higher as more educated population groups are considered. Thus, average satisfaction of life (on a scale of 0 to 10) is 7.8 among those with a primary level of education, 8.0 with secondary education, 8.2 for those who completed secondary school, 8.4 among those with a bachelor's degree, and 8.7 for those who have graduate level studies. The satisfaction with life reported by unmarried people is practically the same as that of married women (8.1 in both cases), while those living as unmarried couples are slightly lower than 8.0, with more significantly lower values for divorcees and widowers (7.7 in both cases) and those currently in a process of separation (7.6).

In terms of happiness, the average for married people is 8.5, slightly higher than those unmarried couples living together as well as singles (8.4 in both cases). They are followed by divorced individuals with 8.1 and somewhat lower (7.9) for separated and widowed individuals. It was also observed that the average levels of life satisfaction for individuals increase for those living in households with higher current per capita expenditures. By dividing the population into quintiles according to their current per capita expenditures, it was observed that life satisfaction in the quintile with the lowest expenditure is 7.6 while that of the quintile with the highest expenditure is 8.5 (INEGI 2012a).

As can be readily seen, Mexicans are objectively found to have poor economic, health, education, and other social aspects. Yet, life satisfaction and happiness is found in abundance. So what we have is a collapse of traditional ideas about well-being and, in turn, a permanent process of change in the social, economic and political conditions. This allows us today to present a fragmented picture of a liquid and fleeting well-being. At the same time, we find new challenges for finding the most appropriate way to measure the development and progress of societies.

These paradoxical results show, on the one hand, high well-being values regarding non-material aspects such as family relationships, autonomy and affective life, and low well-being values across more traditional indicators that include education and economic factors. So that in order to properly understand what is happening, we must necessarily combine both aspects that might help us obtain a better radiographic analysis of the subjective aspects. This amounts to modifying the way we understand development and progress. The various indicators that have been constructed so far fail to integrate these new factors distributed across the diversity of societies around the globe.

In Search of Well-Being

Through a multivariate analysis, the dimensionality of the data can be reduced to describe the values in a smaller subset of variables, although at the expense of some loss of information. This analysis was performed through the application of multifactorial analysis with the method of principal components and rotation with Varimax normalization and Kaiser applied to the set of 173 variables from a self-reported subjective well-being database.[6] The study was carried out among the 18 to 70 year-old population living in Mexico, randomly selected (one per household) within the dwellings that were part of the sample of the National Survey of Household Expenditures during the first quarter of 2012. In total, questionnaires were filled out and collected from 10,654 people distributed throughout the national territory. This was done in order to represent the larger population with a smaller number of variables constructed as linear combinations from the original data.[7]

---

6   The analysis of principal components is the approach pioneered by Hotelling in 1933 although its origins can be found in the orthogonal adjustments by minimum Pearson squares. This has a double utility: 1) It allows for optimal representation in a single, small dimensional space observations of a general p-dimensional space. In this sense, principal components is the first step for identifying the possible latent or non-observed variables that the data generates; and 2) It permits the transformation of original variables correlated in general, into new, uncorrelated variables that facilitate the interpretation of data (Hair, et al. 1999).

7   Data obtained from INEGI, "Bienestar subjetivo. Microdatos," INEGI, http://www3.inegi.org
.mx/sistemas/microdatos/encuestas.aspx?c=34524&s=est.

After determining the main components through the rotated matrix, a graph analysis was constructed using social network theory, processed by three software programs for analysis and visualization of large networks, in addition to using mapping and clustering techniques. The programs used were Ucinet6, Pajek 4 and finally the VOSviewer software. In addition, factorial scores were obtained in order to perform a cluster analysis.[8] As a result of the principal components of subjective well-being, 49 components were obtained and when they were incorporated into network theory, various maps were obtained that show more clearly the main aspects of the comprehensive well-being of Mexicans.[9]

In order to study the relationships between a series of elements, we opted to use new tools that focused at the onset upon relationships. Social network analysis places the emphasis on the study of the relationships as defined between a series of elements (people, groups, organizations, countries, events, and as we would like to add, concepts). Unlike traditional analyses that explain, for example, behaviour in terms of social class and profession, social network analysis focuses on the relationships rather than just the attributes of the various elements (Quiroga 2003). The particularity of this specific type of analysis is that while it emphasizes relationships and their relational properties, it still allows for the attributive aspects to be incorporated.[10]

---

8   These software programs are available on the Internet: Pajek. Program for Large Network Analysis Ver. 4.04, Universidad de Ljubljana, Ljubljana, Slovenia; S.P. Borgatti, Everett M.G., and Freeman L.C., *Ucinet 6 for Windows: Software for Social Network Analysis* (Harvard: Analytic Technologies, 2002); and finally, VOSviewer ver. 1.5.7, Centre for Science and Technology Studies (CWTS) of Leiden University.

9   The database contains the national level results of the Self-Reported Well-being Module as applied to people between the ages of 18 and 70 (one by housing from within housing complexes from the ENGASTO sample) during the period of January and March of 2012 in both urban and rural areas. It includes 10,654 respondents (5,967 corresponding to women and 4,687 to men) with 201 fields that include information about their self-evaluation of their quality of life (on a sale 0–10), of their satisfaction with their life, of their happiness and qualification, of their mood the day prior to completing the instrument, as well as their socio-demographic and socio-economic background. The data bank is offered by INEGI (2012).

10   Prior to social network analysis, the methodological approach most utilized in Sociology was the more functionalist approach on attributes. In that method, the various social actors being studied are characterized based on their specific attributes as they are ordered by distinct variables (e.g., gender, age, political tendency, etc.). The actors constitute the basic unit of analysis and their variable characteristics can be subjected to a statistical and algebraic analysis, serving as a means for classifying actors and interpret-

The analysis of social networks is a scientific method of analysis of great use for ascertaining the patterns of relationships established within the underlying social structure of Mexicans. While embryonic in the 1930s, it later acquired greater significance with the development of cybernetic sciences such that today, it stands among the common methodologies used across the social sciences. In our case, the results of a factorial analysis clearly demonstrated that in order to obtain an adequate measure of well-being, it is necessary to take into account the objective and subjective factors that account for progress and development of a society. This is especially the case if these factors are expected to play a role in sustaining a democracy.

We want to now show all the components and visualize how each of the complex variables are imbedded, and identify each one's main dimensions at the end.

---

ing their actions. This methodological approach began in the Durkheimian tradition with his study of suicide and was later promoted by Lazarsfeld who created a quantitative, representative survey model in the United States (de la Rua 2010).

In Table 10.1, we can see that the most relevant aspects of the first component, *satisfaction with life*, point to affective life, the appearance of people, their social and family life, all as a set of subjective aspects, but at the same time shaped by other more material aspects such as health, the surrounding neighbourhood, their economic situation, and housing conditions. It further includes other aspects such as happiness, overall satisfaction with education received, and aspects linked to assumptions about the future regarding their own personal life achievements, personal security, satisfaction with their work, and global assessments regarding their country. The second component refers to the *frequency of use of computer and Internet, and level of instruction* integrated by contacts via email, enlisting in social media networks and frequency of use of a personal computer. This relation is negative with respect to the level of instruction, signifying that this instructional level does not condition the use of social networks.

TABLE 10.1 *Principal components 1 & 2 of subjective well-being of Mexicans*

Variables	1. Life satisfaction
Can you tell me on a scale of 00 to 10, how satisfied you feel with each one of the following aspects of your life? 9. Your emotional life.	.736
Can you tell me on a scale of 00 to 10, how satisfied you feel with each one of the following aspects of your life? 8. Your appearance.	.726
Can you tell me on a scale of 00 to 10, how satisfied you feel with each one of the following aspects of your life? 10. Your social life.	.710
Can you tell me on a scale of 00 to 10, how satisfied you feel with each one of the following aspects of your life? 6. Your family life.	.668
Can you tell me on a scale of 00 to 10, how satisfied you feel with each one of the following aspects of your life? 7. Your health.	.638
Can you tell me on a scale of 00 to 10, how satisfied you feel with each one of the following aspects of your life? 5. Your neighbourhood or local community.	.610
Can you tell me on a scale of 00 to 10, how satisfied you feel with each one of the following aspects of your life? 2. Your economic situation.	.575
Can you tell me on a scale of 00 to 10, how satisfied you feel with each one of the following aspects of your life? 3. Your housing situation.	.570
On a scale of 00 to 10, how happy are you?	.558
Can you tell me on a scale of 00 to 10, how satisfied you feel with each one of the following aspects of your life? 4. Your educational background.	.557
Can you tell me on a scale of 00 to 10, how satisfied you feel with each one of the following aspects of your life? 4. Your future prospects.	.545

Variables	1. Life satisfaction
On a scale of oo to 10, how satisfied are you with 2. Your life achievements?	.538
Can you tell me on a scale of oo to 10, how satisfied you feel with each one of the following aspects of your life? 11. The country that you live in (Mexico).	.531
On a scale of oo to 10, how satisfied are you with 3. Personal security.	.530
Can you tell me on a scale of oo to 10, how satisfied you feel with each one of the following aspects of your life? 1. Your current work situation.	.400

Variables	2. Level of educational instruction and frequent use of computer and internet
Last week, did you have email contact with ... 2. Friends?	.833
Are you registered in a social media network (facebook, twitter, myspace, sónico, plaxo, linkedIn, etc.) in which you maintain contact and/or express and received opinions?	.790
Do you use your personal computer frequently?	.780
Last week, did you have any email contact with ... 1. Family member who do not live with you?	.760
Educational level?	−.671
Do you have an internet connection at home?	.594
At some point last week, did you ... 2. Read an article be it in a journal or on the internet?	.499
Last week, did you have any telephone contact with ... 2. Friends?	.466
Have you travelled by air at some point in your life?	.344
Do you have a credit card at present?	.342
Last week, did you have any phone contact with ... 1. Family members who do not live with you?	.281

Note: Included in this table and those that follow are mostly the variables with the largest factorial charges, i.e., greater than .4, although some of the variables with lesser values were left to better visualize the larger picture.

Table 10.2 shows that one aspect of life satisfaction is related to happiness and to the situation in the context of the application of the questionnaire. This supports the considerations about the differentiated explanations between happiness and satisfaction with life, conceptions that are linked but differentiated by the valuations that people make over time. Those aspects integrated with component four regarding interpersonal reflection and leisurely pursuits like watching a documentary about history or science, listening to a program of debate about the reality of the country, etc., reveals a component of environmental attitude and even one of protection of pets. This integrates aspects such as avoiding practices like burning or inappropriately disposing of garbage on public roads, trying to use as few plastic bags as possible, separating the garbage for recycling, and doing something to prevent the abuse, suffering or cruelty to animals. The developmental trend of a society is linked to public services that assume forming commitments to the protection of nature and with the possibility for people to enjoy moments of reflection regarding things deemed to be important.

TABLE 10.2 *Principal components 3–5 of subjective well-being of Mexicans*

Variables	3. Happiness
On a scale of 00 to 10, in general, how satisfied are you with your life?	.783
Life satisfaction.	.776
In general, how happy did you feel yesterday?	.632
In general, how tranquil did you feel yesterday?	.575
Happiness.	.548

Variables	4. Reflected in a serious manner during the past week about life and important things
At some point last week, did you ... 15. Meditate or serenely reflect on your life, your family, your country or the world?	.665
At some point last week, did you ... 16. Have a nice talk or conversation with someone about things that are important in one's life?	.619
At some point last week, did you ... 12. See a TV documentary about history, science, discoveries, art, offices, technology, nature or travel?	.530

Variables	4. Reflected in a serious manner during the past week about life and important things
At some point last week, did you ... 11. See or listen to a program of debate or discussion concerning the reality of the country and the world?	.504
At some point last week, did you ... 4. Listen to music while concentrating on the lyrics?	.397
At some point last week, did you ... 14. Play a game of chess or Chinese checkers?	.360
At some point last week, did you ... 10. Go to the theatre or see a movie in which you saw actors engage important issues?	.315
At some point last week, did you ... 3. Read the newspaper?	.294
At some point last week, did you ... 1. Read a book?	.286
Would you say that you have had serious difficulties, big setbacks or adversities during the course of your lifetime?	.213

Variables	5. Green ecological attitude and protection of domestic pets
In the last 12 months, have you ... 2. Avoided throwing trash in a public street or in open spaces?	.688
In the last 12 months, have you ... 3. Avoided burning trash or refuse?	.650
In the last 12 months, have you ... 5. Avoided discarding used batteries with the rest of the garbage?	.630
In the last 12 months, have you ... 6. Attempted to use the least amount of plastic bags possible or use biodegradable bags?	.590
In the last 12 months, have you ... 1. Separated organic from non-organic waste?	.464
In the last 12 months, have you ... 4. Done something in order to prevent abuse, suffering and/or cruelty to animals?	.363

Table 10.3 shows how principal component six maintains correlations with aspects related to the real capacity of people to obtain public services and other services such as education. This aspect is vital for a high rating of social development and not only because of the capacity of the regime to provide them but also because of the real capacity of the population to access public services and education, and above all because of the social cohesion existing in a society to allow their access. On the other hand, an important component of the level of development remains the existing level of discrimination and abuse by social class linked to age, skin colour, gender or physical differences. One other component is giving or having the capacity to support other people through care or personal attention, affection or through financial support. The indicators on civil or non-governmental organization affiliations, alumni and / or self-improvement or self-help organizations all allow for greater social cohesion of a society. However, the variables of social class status and geographic location remain inescapably important when evaluating well-being.

TABLE 10.3   *Principal components 6–10 of subjective well-being of Mexicans*

Variables	6. Have solicited assistance in order to pay bills, school fees, etc.
Over the course of the last three months, have you or has someone else asked to borrow money or requested financial assistance in order to pay ... 4. The electricity, gas or telephone bill?	.708
Over the course of the last three months, have you or has someone else asked to borrow money or requested financial assistance in order to pay ... 3. The water bill?	.686
Over the course of the last three months, have you or has someone else asked to borrow money or requested financial assistance in order to pay ... 1. For foodstuffs?	.650
Over the course of the last three months, have you or has someone else asked to borrow money or requested financial assistance in order to pay ... 6. For medicine or medical services?	.549
Over the course of the last three months, have you or has someone else asked to borrow money or requested financial assistance in order to pay ... 5. For school uniforms or school supplies?	.510

Variables	6. Have solicited assistance in order to pay bills, school fees, etc.
Over the course of the last three months, have you or has someone else asked to borrow money or requested financial assistance in order to pay ... 2. The monthly rent?	.421

Variables	7. Mistreated for reason of social class, physical traits or gender
Being in Mexico, have you ever been mistreated simply on account of: 10. Your social class standing?	.632
Being in Mexico, have you ever been mistreated simply on account of: 1. Age? (for being young or not being so)	.580
Being in Mexico, have you ever been mistreated simply on account of: 2. The colour of your skin or ethnicity?	.574
Being in Mexico, have you ever been mistreated simply on account of: 9. Defects in your physical body?	.524
Being in Mexico, have you ever been mistreated simply on account of: 4. Your sex?	.493

Variables	8. Someone in your life has required your assistance
Presently, is there somebody in your life that needs ... 1. Your attention or care?	.802
Presently, is there somebody in your life that needs ... 2. Your emotional support?	.787
Presently, is there somebody in your life that needs ... 3. Your economic support?	.746

TABLE 10.3 *Principal components 6–10 of subjective well-being of Mexicans* (cont.)

Variables	9. Membership in a civic organization, NGO, etc.
Do you belong to a ... 8. Non-governmental organization (NGO)?	.662
Do you belong to a ... 9. Voluntary or philanthropic association?	.653
Do you belong to a ... 12. Any other civic association (Scouts, Rotary Clubs, Lion's Club ...)?	.522
Do you belong to a ... 5. Student or alumni organization?	.346
Do you belong to a ... 10. A self-help group and/or group for coping with personal problems?	.321

Variables	10. Social status and indigenous language
Distribution stratum	−.611
CONAPO stratum	.557
Do you speak any of the following languages? 1. Original language of Mexico? (e.g., náhuatl, maya, mixe, otomí, tarasco, etc.)	.476
State of the Republic	−.347

In Table 10.4, an important aspect to emphasize is the capacity for people to economically support their families. This generates greater self-esteem and freedom for people to make decisions, in addition to their having experienced happiness in childhood and adolescence that maintains personal continuity with the present moment and their expectations for the future. The following components are linked to the identification with a sports organization, the availability of free time, and the achievements and recognitions that people have received for their efforts and freely made decisions.

TABLE 10.4 *Principal components 11–17 of subjective well-being of Mexicans*

Variables	11. Have you given economic assistance to family members or friends
Over the last 12 months, have you ... 1. Economically supported family members or relatives that live in another household?	.668
Over the last 12 months, have you ... 2. Economically assisted or have you otherwise assisted people who are not relatives?	.581

Variables	12. Experienced happiness in childhood or adolescence
Have you experienced a moment of great joy or happiness? 2. As an adolescent.	.812
Have you experienced a moment of great joy or happiness? 1. As a young child.	.807
Have you experienced a moment of great joy or happiness? 3. After the age of 17.	.381

Variables	13. Membership in a sports team
Have you in the last week practiced any sporting activity in which you competed against another person or another team?	.770
Do you belong to a ... 11. League or other sporting association?	.738

TABLE 10.4 *Principal components 11–17 of subjective well-being of Mexicans* (cont.)

Variables	13. Membership in a sports team
Have you in the last week carried out any physical activity for 30 minutes or more such as walking, jogging, swimming, cycling, dancing, yoga, Tai-chi, or gym workout?	.388
Male or female?	.377

Variables	14. Free Time
Do you have any free time during the week?	.635
On a scale of 00 to 10, how satisfied are you with 1. Free time to do things that you like?	−.535
Think for a moment about the activity that you most like to do or which gives you the most pleasure in life ... Have you performed this activity over the last week?	.511

Variables	15. Achievements and personal awards
Do you consider that you have had achievements or have managed to get something that makes you feel good?	.644
Have you ever received a sincere act of gratitude, honorary acknowledgement or recognition for your efforts in doing things well?	.577
Do you consider that over the course of your life, you have been able to make important decisions freely?	.469

Variables	16. Have complained about water leaks and garbage strewn in public space
Over the last 12 monthes, have you ... 10. Filed a complaint against people who throw trash in the street, in parks, rivers or beaches?	.758
Over the last 12 months, have you ... 9. Filed a complaint about water line leaks in a public area?	.735

Variables	17. Anger and sadness experience yesterday
In general, how angry did you feel yesterday?	.836
In general, how sad did you feel yesterday?	.801

Table 10.5 displays the following six components: attendance at concerts or practical instruction classes, disability, having support in case of an emergency, having pets, imagining Mexico's future and its own well-being, and suffering threats or aggressions from someone close with whom they live. The first components are strongly correlated with attendance at cooking classes, weaving or handicrafts; disability is mainly correlated with the permanent use of crutches, wheelchairs, a walker, a walking stick, or a prosthesis in legs or arms, suggesting that this component is linked to old age and situations caused by accidents. The number of people in the family with whom a person can count on in case of an emergency is an important aspect of the social support component. The image that people have about the situation of the country for the upcoming ten years is correlated with personal situations in terms of economic well-being.

TABLE 10.5   *Principal components 18–23 of subjective well-being of Mexicans*

Variables	18. Attended a concert, or handicrafts, shop, or cooking classes
At some moment in the past week, have you ... 7. Attended a cooking class, a knitting class or some other handicraft skill class?	.571
At some moment in the past week, have you ... 5. Attended a dance, painting, music, gardening or photography class?	.566
At some moment in the past week, have you ... 9. Gone to a concert or a musical performance?	.509
At some moment in the past week, have you ... 13. Attended a conference, visited a museum, art gallery or an exposition?	.381
At some moment in the past week, have you ... 8. sung or played a musical instrument?	.315

Variables	19. Handicapped
Do you use crutches, wheel chair, support, cane, a leg or arm prosthesis, or require any other type of assistance to move on a permanent basis?	.690
Do you have a physical problem or difficulty in hearing, and/or verbally communicating?	.610

Variables	19. Handicapped
Being in Mexico, have you ever been mistreated simply on account of being handicapped or having a physical impediment?	.508
At present, are you suffering the consequences of an accident or serious illness that requires medical care for more than a one year period?	.485

Variables	20. Have persons that can be counted on in the event of a problem
How many people in your family can you count on in the event of an emergency or situation requiring urgent assistance?	.714
How many people who are not members of your family can you count on in the event of an emergency or situation requiring urgent assistance?	.694
How many neighbours know you by your name?	.442

Variables	21. Have pet
Do you have a pet or domesticated animal that helps keep you company?	.640
Over the last 12 months, have you ... 7. Planted a tree?	.484
Over the last 12 months, have you ... 8. Cared for plants, trees or flowers in your house or neighbourhood?	.440

Variables	22. Have imagined Mexico's future and its impact on personal economic well-being
How do you see yourself in Mexico over the next ten years?	.820
How do you see yourself in terms of economic well-being over the next 10 years?	.746

TABLE 10.5   *Principal components 18–23 of subjective well-being of Mexicans* (cont.)

Variables	23. Have been threatened or have suffered personal harm by someone living in the same household
During the last 12 months, have you received threats from ... 1. Someone whom you live with?	.826
During the last 12 months, have you suffered or been the object of physical aggression on the part of ... 1. Someone whom you live with?	.816

In Table 10.6 we can find the following components such as religious affiliation having to do with the active promotion of a faith and/or religious values, and donations to cultural organizations or those that help other people, especially if involving donations to institutions that promote the conservation of culture, nature or urban spaces. The component of strength and self-confidence is mainly defined by the scale of dependence and personal capacity to face the adversities of life and to personally do well, and how to influence the situation that the country maintains that affects their own well-being. Two more components involve: consideration about having better opportunities than their parents for education and work; and the capacity to speak a foreign language.

TABLE 10.6   *Principal components 24–28 of subjective well-being of Mexicans*

Variables	24. Active religious affiliation
Do you belong to a ... 2. Group or association in which you actively promote faith or religious values?	.694
Do you belong to a 1. Church or religion?	.530
Being in Mexico, have you ever been mistreated simply for ... 7. Religious reasons?	.404

Variables	25. Have made donation to a cultural organization or to help other people
During the last 12 months, have you ... 4. Made donations to an institution that promotes care or conservation of culture, nature or urban space?	.722
During the last 12 months, have you ... 43. Made donations (not to beggars) to an organization dedicated to helping people?	.677
During the last 12 months, have you ... 45. Carried out some kind of voluntary or community work?	.413

TABLE 10.6 *Principal components 24–28 of subjective well-being of Mexicans* (cont.)

Variables	26. Conviction and confidence in themselves
On a scale from oo to 1o, how much do things depend on you that this year and next goes well?	.509
On a scale from oo to 1o, how strong do you consider yourself in the face of life's adversities?	.463
On a scale from oo to 1o, how much does the country's situation influence your well-being?	.317
Normally, how often to you meet with family members who do not reside in your household?	.292

Variables	27. Consider that they have had better education and work opportunities than their parents
Do you consider that you have had better work opportunities in Mexico than did your parents or the people who raised you?	.671
Do you consider that you have had better educational opportunities than did your parents or the people who raised you?	.570
Do you consider that you have had better opportunities to acquire patrimony (house, apartment, land) than did your parents or the people who raised you?	.433

Variables	28. Speak a foreign language
Do you speak any of the following languages? 3. French, Japanese, or other language other than Spanish?	.663
Do you speak any of the following languages? 2. English (conversational)	.589
At some point in the last week, have you ... 6. Studied another language on your own, in a language institute, or with a private instructor?	.448

Table 10.7 shows the components most related to the kinds of specific situations that Mexicans experience and which can be integrated into the composite indicator. This includes items such as levels of alcoholism or drug addiction; having a relative suffering some consequence of an accident; receiving threats or suffering physical attacks by someone known to the individual; affliction by a serious physical or mental illness; threats by unknown people; three components that we found linked to educational attainment; the possibility of making decisions freely; and the number of people who contribute income to the household.

TABLE 10.7 *Principal components 29–36 of subjective well-being of Mexicans*

Variables	29. Sharing life with someone with drug use or alcohol problems
Do any of the people whom you live with suffer from ... 2. Drug addiction?	.696
Do any of the people whom you live with suffer from ... 1. Alcoholism?	.602
Do any of the people whom you live with suffer from ... 1. Incarceration?	.499
Do any of the people whom you live with suffer from ... 2. Missing or disappeared?	.406

Variables	30. Someone in family suffered consequences of an accident
Presently, do any of the following people in your life suffer from the consequences of a severe accident who will not be able to fully recover? 5. Grandchild.	.758
Presently, do any of the following people in your life suffer from a serious illness? 5. Grandchild.	.642
Presently, do any of the following people in your life suffer from the consequences of a severe accident who will not be able to fully recover? 4. Child.	.477

TABLE 10.7 *Principal components 29–36 of subjective well-being of Mexicans* (cont.)

Variables	31. Object of threats or physical aggression
During the last 12 months, have you received threats from ... 2. Another person that you know?	.784
During the last 12 months, do you suffer or have you been the object of physical aggression from ... 2. Another person that you know?	.768

Variables	32. Educational level
Years of education?	.761
Final year of education completed?	−.728

Variables	33. Someone close suffering from physical or mental illness
Presently, do any of the following people in your life suffer from a grave or debilitating illness? 4. Child.	.697
Do any of the people with whom you live suffer from ... 3. A physical or mental illness?	.633

Variables	34. Possibilities for taking decisions freely
Do you consider that the possibility of making decisions freely has been present?	.693
How does the standard of living in your present household compare to the one in which you grew up?	.498
When was the last time that you received recognition or gratitude for your efforts?	.244

Variables	35. Threats or aggression by some unknown person
During the last 12 months, do you suffer or have you been the object of physical aggression from ... 3. An unknown person?	.762
During the last 12 months, have you received threats from ... 3. Another unknown person?	.761

Variables	36. Persons in the household who contribute income
How many people in your household contribute income at this moment?	.701
Number of people in your household?	.656

Components 37 to 45 were generally integrated with fewer variables. This may indicate importance in variables themselves, such as the frequency of meetings with friends and the serious illness of a sibling, intimate partner or their parents. Two components were related to abuse, namely, on account of political preference and sexual preference (Table 10.8). Components related to belonging to parental organizations or neighbourhood organizations were seen to relate to aspects of social cohesion.

TABLE 10.8   *Principal components 37–45 of subjective well-being of Mexicans*

Variables	37. Frequency of meeting with friends
Normally, how often do you meet with your friends?	.132

Variables	38. Severe illness or accident of a brother or sister
Presently, do any of the following people in your life suffer from the consequences of a severe accident who will not be able to fully recover? 3. Sibling.	.760
Presently, do any of the following people in your life suffer from a serious illness? 3. Sibling.	.716

Variables	39. Mistreatment on account of political preferences
Being in Mexico, have you ever been mistreated simply on account of political preferences?	.464
Do you belong to ... 3. A party, movement, or another political or social organization?	.459
16. Do you belong to ... 4. A professional organization, occupational association, or trade union?	.357

Variables	40. Severe illness of a parent
Presently, do any of the following people in your life suffer from the consequences of a severe accident who will not be able to fully recover? 2. Parent.	.748
Presently, do any of the following people in your life suffer from a serious illness? 2. Parent.	.702

Variables	41. Affiliation with a parents or neighbourhood organization
Do you belong to ... 6. The board of a parents association?	.661
Do you belong to ... 7. A neighbourhood association?	.539

Variables	42. Age of the greatest adversities
At what point of your life did you face the greatest adversities?	.532
Age.	.320
Has someone close to you died during the last ten years?	−.271

Variables	43. Severe accident or illness of your life partners
Presently, do any of the following people in your life suffer from the consequences of a severe accident who will not be able to fully recover? 1. Conjugal partner.	.733
Presently, do any of the following people in your life suffer from a serious illness? 1. Conjugal partner.	.637

Variables	44. Quality of sleep
How well did you sleep for the majority of times during the last week?	−.707
How well did you sleep last night or at the hour that you should normally sleep?	−.409

TABLE 10.8   *Principal components 37–45 of subjective well-being of Mexicans* (cont.)

Variables	45. Mistreated on account of sexual preference
Being in Mexico, have you ever been mistreated simply on account of ... 5. Sexual preference or sexual orientation?	.668
Being in Mexico, have you ever been mistreated simply on account of ... 12. Another motive?	−.436

The last components (Table 10.9) shows how the development of a society is affected after a death occurs of a close person. Two aspects related to abuse are by a stranger and by having AIDS and, finally, by the frequency of physical activity and marital status. It is important to analyse all of the components in order to visualize their integration in a smaller number of aspects that are obtained in the cluster analyses and networks.

TABLE 10.9   *Principal components 46–49 of subjective well-being of Mexicans*

Variables	46. Amount of time passed since a closely related deceased person
How long has it been since the death of someone who most affected you?	.729

Variables	47. Mistreated for being a foreigner
Being in Mexico, have you ever been mistreated simply on account of ... 11. Being a foreigner?	.670

Variables	48. Mistreated for living with HIV/AIDS
Being in Mexico, have you ever been mistreated simply on account of ... 6. Having AIDS?	.660

Variables	49. Frequency of physical activity and marital status
How many times have you performed physical activity during the past week?	.712
Marital status.	−.287

From the social networks perspective, a multifactorial analysis was performed to produce the following relational maps that show the main components of Mexican well-being and their relationships. The algorithms for constructing the maps can be read in terms of the distance between points or aggregates of points, and by the number of relationships and outstanding relationships. In empirical terms, they would be the existing relationships between the people surveyed and in that sense we can see the global trends of what is within the subjectivity of people as inherent aspects of a political regime in which they coexist.

Figure 10.1 and 10.2 show the 24 main components integrated in the network of relationships where what stands out are the components: satisfaction with life; frequent use of the Internet and computers with level of instruction; and ecological, green attitude and the protection of pets. Satisfaction with life closely relates in positive terms to religious affiliation, seeing and hearing debate programs or discussions about the state of the country, and whether people have provided financial support to relatives or acquaintances. The same components have a negative relationship with social class abuse, political preferences, living with AIDS; i.e. by social discrimination, along with suffering from alcoholism or drug addiction.

The frequent use of computers and the Internet as well as educational level is the closest to the age of greatest adversities and to the number of people in the household who contribute income, while on the other hand, ecological attitude is linked to the imagination of people in terms of their economic well-being and the image they have of their country and its prospects for the coming years.

The visualization offered in Figure 10.3 in multiple levels of clusters in a frame format offers a more focused look at the relational factors. If we observe them as a Cartesian plane, in the centre we have religious affiliation as an important aspect close to life satisfaction in the lower part. We can also see how the negative factors group together such as social abuse, alcoholism, disability, the suffering of a family member from the consequences of an accident or suffering from a physical impairment that makes it difficult to listen or communicate verbally. In positive terms, after the indicator of frequency in the use of computers and the Internet, we find the number of people who actively contribute income to the home, the personal achievements and recognition of people, an ecological attitude, the self-image people have about their economic well-being, the image they have of their country, having had success in personal achievements that gives a good feeling, and having experienced moments of joy or happiness as a teenager. In empirical terms, we have obtained a radiography of the most important aspects that define well-being.

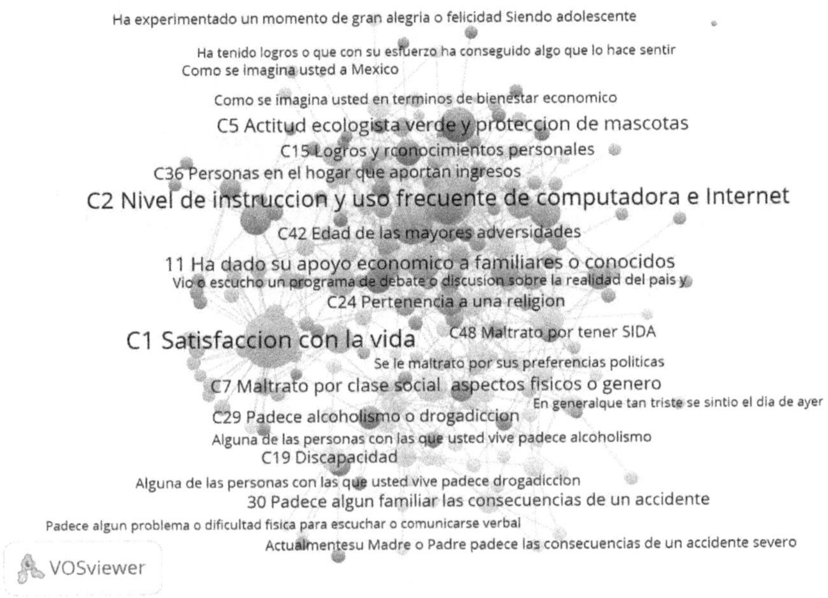

Ha experimentado un momento de gran alegria o felicidad Siendo adolescente

Ha tenido logros o que con su esfuerzo ha conseguido algo que lo hace sentir
Como se imagina usted a Mexico

Como se imagina usted en terminos de bienestar economico

C5 Actitud ecologista verde y proteccion de mascotas

C15 Logros y rconocimientos personales

C36 Personas en el hogar que aportan ingresos

C2 Nivel de instruccion y uso frecuente de computadora e Internet

C42 Edad de las mayores adversidades

11 Ha dado su apoyo economico a familiares o conocidos
Vio o escucho un programa de debate o discusion sobre la realidad del pais y
C24 Pertenencia a una religion

C1 Satisfaccion con la vida   C48 Maltrato por tener SIDA

Se le maltrato por sus preferencias politicas

C7 Maltrato por clase social aspectos fisicos o genero
En generalque tan triste se sintio el dia de ayer
C29 Padece alcoholismo o drogadiccion

Alguna de las personas con las que usted vive padece alcoholismo
C19 Discapacidad

Alguna de las personas con las que usted vive padece drogadiccion
30 Padece algun familiar las consecuencias de un accidente

Padece algun problema o dificultad fisica para escuchar o comunicarse verbal
Actualmentesu Madre o Padre padece las consecuencias de un accidente severo

FIGURE 10.1   *Multi-level clustering of subjective well-being of Mexicans*

Ha experimentado un momento de gran alegria o felicidad Siendo adolescente

Ha tenido logros o que con su esfuerzo ha conseguido algo que lo hace sentir
Como se imagina usted a Mexico

Como se imagina usted en terminos de bienestar economico

C5 Actitud ecologista verde y proteccion de mascotas

C15 Logros y rconocimientos personales

C36 Personas en el hogar que aportan ingresos

C2 Nivel de instruccion y uso frecuente de computadora e Internet

C42 Edad de las mayores adversidades

11 Ha dado su apoyo economico a familiares o conocidos
Vio o escucho un programa de debate o discusion sobre la realidad del pais y
C24 Pertenencia a una religion

C1 Satisfaccion con la vida   C48 Maltrato por tener SIDA

Se le maltrato por sus preferencias politicas

C7 Maltrato por clase social aspectos fisicos o genero
En generalque tan triste se sintio el dia de ayer
C29 Padece alcoholismo o drogadiccion

Alguna de las personas con las que usted vive padece alcoholismo
C19 Discapacidad

Alguna de las personas con las que usted vive padece drogadiccion
30 Padece algun familiar las consecuencias de un accidente

Padece algun problema o dificultad fisica para escuchar o comunicarse verbal
Actualmentesu Madre o Padre padece las consecuencias de un accidente severo

VOSviewer

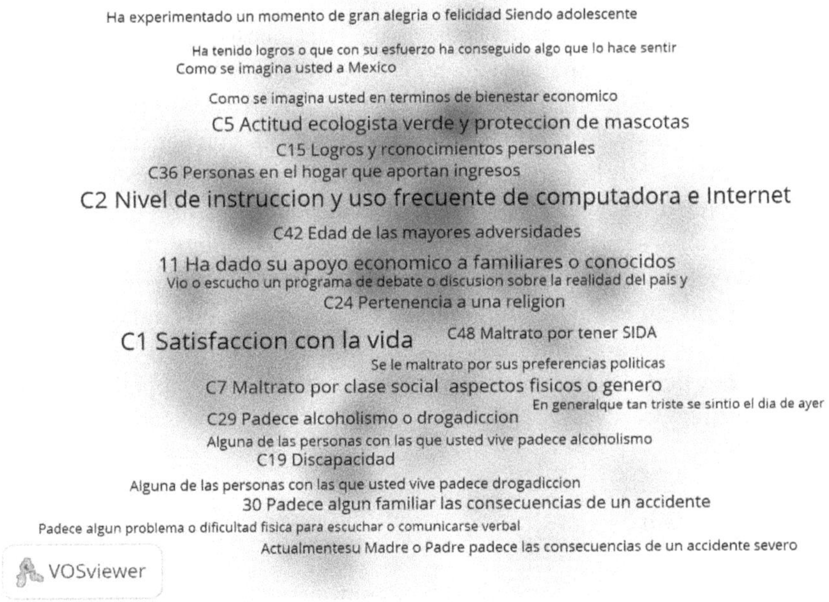

FIGURE 10.2   *Network visualization of multi-level clustering of subjective well-being of
Mexicans*

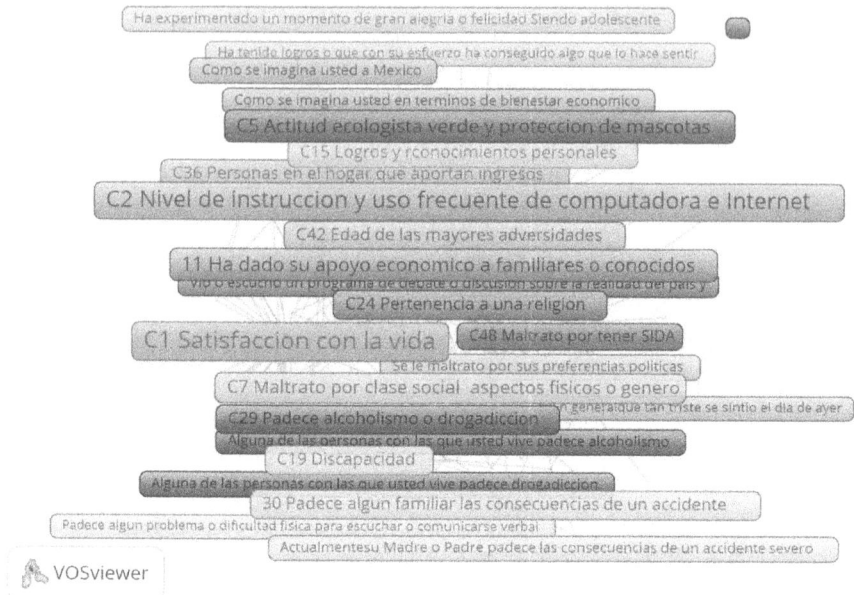

FIGURE 10.3   *Network frames visualization of multi-level clustering of subjective well-being of Mexicans*

Usually, measures of centrality speak to analysis focused on specific sectors of the network so as to analyse which of the points may have more capacity or influence, or which are necessary to support the structure of relations at the time of the study. In this sense, Figure 10.4 shows that satisfaction with life is sustained mainly by satisfaction with social life, events of happiness, and having had difficulties, setbacks or adversities during their lives.

Figure 10.5 shows structural holes that identify which points of the network are necessary so that the structure does not disintegrate. What we observe is that life satisfaction has to be supported by self-effort and self-confidence, and also whether people are sometimes mistreated because of the colour of their skin, or if they derive satisfaction from their social life, their appearance and their family life.

Conclusion

What was obtained in the end is a structural complex/index (though not structural in a rigid, objective sense) of the main empirical characteristics of well-being, composed of complex indicators and constructed by examining

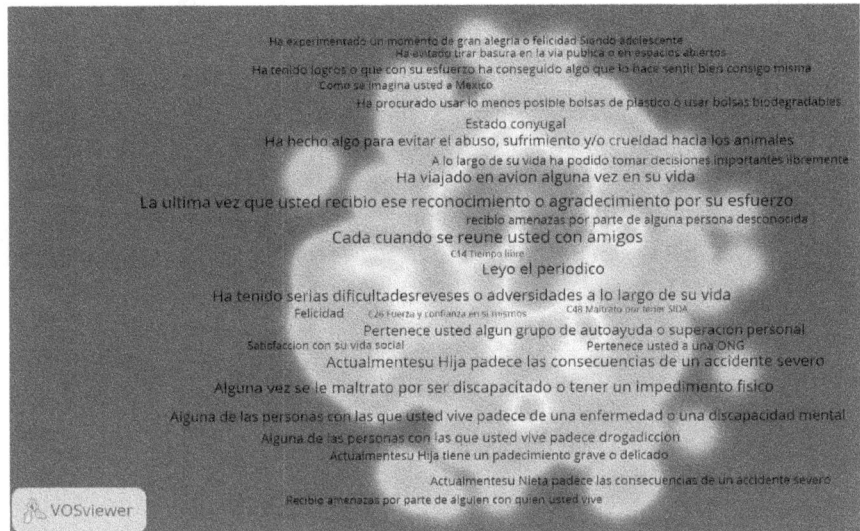

FIGURE 10.4   *Output degree centralization of subjective well-being of Mexicans (network output degree centralization = 0.03425638)*

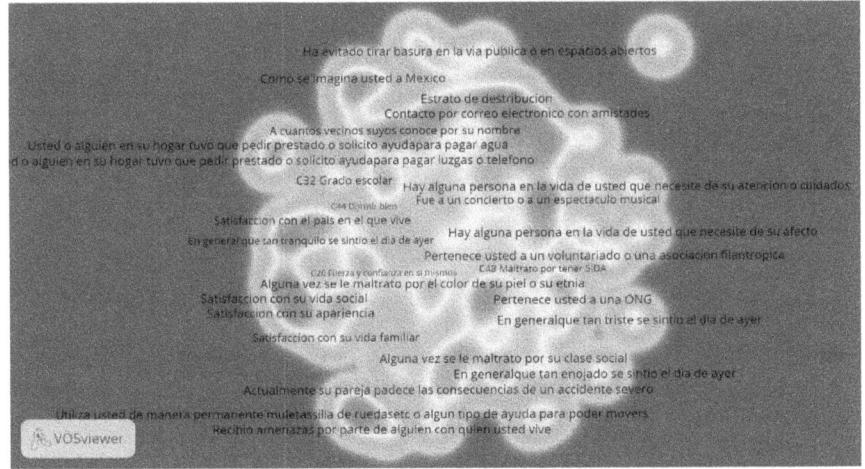

FIGURE 10.5   *Structural holes in the subjective well-being of Mexicans*

the matrices that register the relationships between actors/concepts. Through their density, connectivity, segmentation or classes of structural equivalence, distances, centrality, etc., we can describe the underlying tendencies of the subjective aspects that go beyond the traditional indicators of societal development. It can be theoretically posited that the establishment of those relationships or the impossibility to create them depend upon the way in which the existing relations of production are organized.

Thus we can conclude that the network/map constitutes a relational system, a more theoretical concept than just the network, and we can hypothesize about the interdependence of relations between the constituent components. What we call structure amounts to the principal rules that ensure the production of this system. In this manner, we propose and highlight the notion that subjective well-being can be integrally placed as a key element for considering a newer way of measuring the development of a society.

CHAPTER 11

The Challenges of Democracy in Mexico

Héctor de la Fuente Limón

This study analyses some of the challenges facing democracy in Mexico, based on the neoliberal form of accumulation imposed on the country in recent years. The effects of this imposition can be seen in an economy with growth rates lower than the needs of an increasing population of working age. It is an economy lacking the structural capacity to generate employment, with abrupt and steady declines in real wages, increasingly precarious employment, increasing social inequality and economic exclusion, increasing informality, and increased migration as a survival strategy alongside of the growth of an illegal economy and social violence.

Under these kinds of conditions, a set of social uncertainties and conflicts has been generated that have restricted the broad and equal participation of citizens in public affairs, undermined the autonomous forms of collective organization, and restricted the rule of law. The type of relationships that have been established between the state and society have become clientelistic, vertical, and increasingly violent relations that the society reproduces and manifests back to the state. The crisis of legitimacy that pervades established parties and the main democratic institutions demonstrates a clear distancing with a political regime that increasingly governs behind the backs of the citizens. All these elements allow us to reflect, in a broader sense, on the limits imposed upon democracy by underdevelopment, generating a set of inequalities and social conflicts that limit it and make it unfeasible.

In the first section, some conceptual elements regarding politics, the state, and democracy are outlined, focusing on an analysis of the negative changes that have been experienced in recent years in the context of the neoliberal globalization process. The following section provides an overview of the situation of the Mexican economy in recent decades, highlighting its exclusionary, highly exploitative, and unequal organization. Section three analyses the crisis of representation caused by the social inequalities generated by the economic organization of the country, particularly with respect to the crisis of political parties, the existence of a limited and unequal citizenship, a crisis in the rule of law, and the proliferation of criminality in all spheres of social life. Some final reflections in light of the above are made at the end.

Transformations of Democracy under Neoliberalism

Politics are imbued with social relations of domination that regulate social antagonisms that result from class, race, ethnicity, and gender divisions within and between social formations. The principal agent of politics in capitalist society is the state and as such, its organization and functions are determined by the historical conditions of domination. The form that the organization and the functions of the state acquire in order to regulate political conflict in a determined historical context constitutes a political regime. That is to say, it represents the type of relations of domination that are established between the state and civil society, and they determine the possibilities that the society may have for influencing state decision-making, and in the final analysis, it defines how the social surplus is distributed. The regime embodies politics and the state embodies power (González Casanova 1992).

The regulation of political conflict within countries in the post-WWII period took on distinct forms whose boundaries have been marked at one end by democratic regimes and on the other by various types of authoritarianism. The differences between the two can be located, as Charles Tilly (2007) argues, in the type of relations established between the state and its citizens as based on at least four elements that vary in levels and degrees in a given historical context: 1) political inclusion of the population under the jurisdiction of the state, through the recognition of rights and civil liberties; 2) equality between and within different categories of citizens in the recognition of the rights and duties of citizens; 3) protection of the population from arbitrariness of the state; and 4) mutually linked consultation on public policy decisions (Tilly 2007: 45–46).

These elements, in turn, depend on the state's capacity to enforce its political decisions and this implies not only efficacy in the use of its monopoly on violence but above all the legitimacy of those decisions. This capacity is closely linked to room to manoeuvre that the State retains as it intervenes in the organization of the capitalist economy.

We can speak then of two types of regimes in conflict, one that operates according to the dictates expressed in the "free play of market forces" and the other based on social needs or rights as expressed in the collective options of democratic politics as supported by the exercise of political, civil and social rights (Streeck 2011). These regimes are mutually determined and in permanent tension where the state intervenes in order to define the distribution of socially produced material and immaterial goods. Democratically elected governments should attend to both of these principles. With the exception of the period in which the welfare state flourished through achieving a certain balance between the two in some of the more advanced countries of the Western world, the trend over the last thirty years has been to privilege the market regime.

During the rise of the Keynesian pattern of accumulation with its emphasis on maintaining full employment and a sustained aggregate demand as a condition for the stability of capitalism, democracy was consolidated through the welfare state. In this context, workers' organizations exercised broad influence in parliaments and governments, and political parties operated as intermediaries between the state and the demands of civil society for participation and social justice. It was a social pact that guaranteed an active flow of concessions between social classes and societal groups that resulted in highly inclusive and egalitarian regimes in the exercise of citizenship, and protectors of rights and freedoms, bolstered by strong and legitimate governments.

Even in the countries of the capitalist periphery, populist regimes were established early on, which laid the foundations for the implementation of the import-substitution model of industrialization, generating economic growth, substantial employment rates and acceptable levels of wages that have not been equalled at any other time in history. Important support came from the corporatist mechanisms for the mediation of interests between the state and rural and urban workers that effectively extended citizenship by opening the way for the installation of a social state, although limited by some restrictions to political rights and freedoms.

Developmentalism over the long post-war period constituted the continuation of this process, albeit with much greater limitations on the participation of the popular classes in the intermediation of interests, with predominance of national and foreign capital. In this context, Western democracy rather than being a constant in these countries was instead a broad melting pot of regimes occupying an intermediate position between democracy and authoritarianism, albeit imbued with a nationalist and progressive vision.

The imposition of the neoliberal pattern of capitalist accumulation had as its lynchpin the rupture of the social pact of the post-war period. The capitalist crisis that manifested itself in the sixties and seventies required a change in the political and economic organization of the state. In economic terms, the priority was to recover the rate of profit and it achieved this by attacking trade unions, breaking up the welfare state, and implementing a more flexible labour regime.

In this context the mechanisms of the previous period for mediating competing class interests were undermined. Political parties entered into a long period of discredit by being more closely tied to state organs rather than operating as intermediaries in the face of demands from civil society. Citizenship became limited and vastly unequal for many social groups and the rule of law was relegated to being a mechanism for reproducing social inequalities rather than protecting rights from abuses of authority and arbitrariness of the state. As the role of the state became diminished, the door was wide open for large transnational corporations to fully capture the benefits being offered up by the liberalization of the economy and the financing of the world market. Production became increasingly de-territorialized, new technological advances were introduced and a new correlation between social classes was established. Democracy became more an electoral mechanism for the renewal of elites than a space that could house broad, egalitarian, protective and mutually binding relations between the state and civil society, and between social classes as in the previous period.

In the capitalist periphery of which the emblematic case is Latin America, these processes led to the establishment of repressive dictatorships in the political sphere and broadly liberal regimes in the economic sphere.[1] This opened the way for periods of transition to a type of procedural democracy that has since become widely functional in the political regulation of the state for the reproduction of the neoliberal pattern of accumulation. Although strong popular support regimes have been consolidated in the region in recent years that have expanded the conditions for broad and inclusive citizenship, including experiences of greater state intervention in the economy and participation in social policy such as Venezuela, Bolivia and to a lesser extent Ecuador, Argentina, Brazil and Uruguay, the general variables of the neoliberal pattern of accumulation have essentially been maintained.

This demonstrates that the path taken by democracy is easily reversible in social contexts characterized by weak states that are highly sensitive to and even dependent upon the interventionism of foreign governments, agencies and businesses from more developed countries. But it also shows that neoliberalism remains highly incompatible with democracy and the egalitarian, pro-

---

1   The Pinochet Dictatorship in Chile will be remembered as the first laboratory for neoliberalism in the world, implemented by strongly authoritarian means. The incompatibility of this pattern of accumulation with the more inclusive democracy of the preceding period was openly recognized by the theoretical mentor of this economic doctrine, Frederic Hayek, in an interview given in that country in April, 1981 when he stated: "… evidently the dictatorships carry along with them risks. But a dictatorship can be self-limited and these self-imposed limits can be more liberal in their policies than a democratic assembly that has no such limits. A dictatorship can be the only hope and can be the best solution in spite of everything" (Escalante 2015).

tective and consultative mechanisms for public policy formulation that were developed earlier in the post-WWII period. Given the dilemmas involved in the organization of the state to address both the principles of marginal profits and those of social justice, the former has been imposed at the expense of the latter over the course of recent decades.

Mexico is perhaps the best example of these trends. More than thirty years of neoliberal orthodoxy has disarticulated the social pact that underpinned the hegemony of the Institutional Revolutionary Party (PRI) regime. This was characterized by the establishment of authoritarian but pragmatic regulation of economic policy and construction of the social state, limited to salaried urban workers and mediated by a strongly hierarchical corporate trade union structure.

The weak democracy that was built in the period of alternation, based on the recognition of political rights and civil liberties undermined in the authoritarian period, is now up against the ropes. Inequality, poverty and social exclusion have spread widely while citizenship is precarious for large sectors of the population. Political parties increasingly resemble cliques that govern against their constituents and have become widely discredited. The rule of law is disappearing, and the governments that result from elections are increasingly questioned, opaque in their decision making, unaccountable and corrupt. Under these conditions, democracy is not operating as a space for negotiation between social classes around issues of public decision-making. Instead, the spaces of political inclusion have ceased to correspond with the social and economic spheres, leaving formally elected governments devoid of their legitimacy.

In the following sections, the focus is placed on characterizing democracy in Mexico based on the social effects that the political and economic organization of the state has had on the larger society over the neoliberal period. Special evidence is placed on the most recent trends. This analysis will allow us to better appreciate the crisis in the democratic institutions of mediation that has intensified in recent years, along with the challenges that democracy faces in order to reconstitute itself on new foundations.

The Social Effects of the Neoliberal Pattern of Accumulation

The restructuring of the world economy that intensified in the 1980s has had very strong and lasting effects on conditions of reproduction and social class relations. The strategies that were implemented during that period in order to overcome the crisis, i.e., liberalization of economies, financialization of markets, and increased labour flexibility in productive processes, formed the back-bone of a globalization that acquired the form of accumulation by dispossession, or better put, accumulation as a permanent process (Harvey 2007).

In Mexico, the labour force expanded during this period as in no other period in history. It did so in the midst of a process that involved the separation of thousands of people from their lands, their jobs and their most basic rights. Capitalist restructuring has been based primarily on the objective to put an end to self-sufficient economies and the means for achieving them in order to relaunch the process of accumulation alongside of reduced labour costs.

In broad terms, the growth of the economy and the distribution of income are indicators of the levels of well-being in a population. In Mexico, average GDP growth rate reached during the neoliberal period has been around one-third or less (2.2%) than the average of 7.1% achieved during the period of industrialization under the import substitution strategy between 1950 and 1980 (Calva 2005). This weak economic expansion has had a strong impact on the generation of formal employment in recent decades. If this indicator is measured by the variation in the number of permanently insured persons covered by the three main sources of insurance (IMSS, ISSSTE and PEMEX) per decade, an abrupt downward trend can be observed and this has been accentuated since the latter 1990s. After a period of accelerated growth in 1960–1970 of 10.8%, the structural crisis of the industrialization model via import substitution saw growth drop by almost half (5.7%) and this trend continued until reaching a growth rate of just 1.3% in the decade 2000–2010 (Samaniego 2015: 14).

This is a product of the conditions under which the country entered the world economy, i.e., as a centre of manufacturing that operates with cheap labour in which workers are hired under some of the worst of conditions. According to Norma Samaniego, a significant growth of employment in this sector took place during the second half of the 1990s thanks to the growth of the *maquila* plants that ultimately employed around four and a half million people. But there was stagnation during the first decade of the 21st Century, leading the number of people employed in *maquilas* to fall until it was at less than 3.5 million in the context of the global crisis of 2009. From that point through 2014, there was a significant increase in the generation of jobs in this sector until reaching in 2014 around the same figures as the first decade of the 1990s. This was motivated to a large extent by the rebound of the automotive industry (Samaniego 2015: 27).

Employment in the countryside experienced an even steeper decline. With the percentage of employment above 25% of the economically active popu-

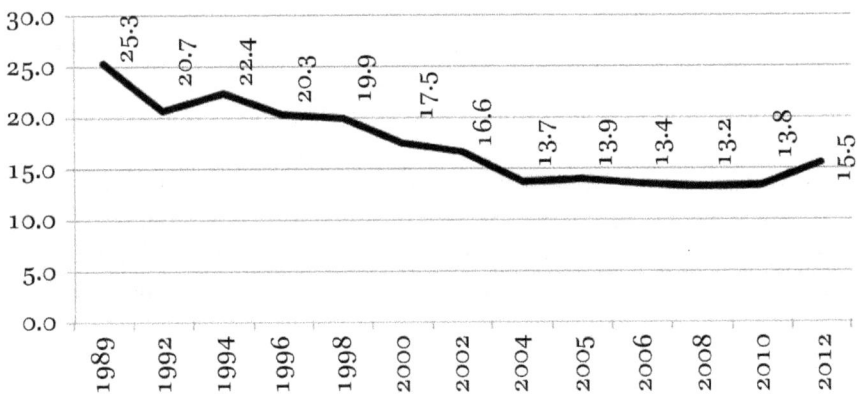

FIGURE 11.1   *Economically active population employed in agriculture, mining, hunting and fishing*
SOURCE: CEPAL DATABASE

lation in 1989, the reform of the cooperatively farmed *ejido* and the entry into force of the North American Free Trade Agreement (NAFTA) led the percentage of employed rural workers to steadily decline until reaching 13.3% in 2010, recovering in 2012 but only to reach 15.5%, largely due to declining migration. (See Figure 11.1)

What we have then is a labour market strangled in the urban environment that suffers the pressure exerted by a growing demand of jobs from displaced rural workers. According to figures from a recent population census, 7 out of 10 rural municipalities in Mexico have experienced depopulation. The traditional relief valves for the entire working-age population unable to find employment have been migration and the informal economy. According to figures from the Mexican National Population Council (CONAPO), the number of international emigrants aged 14 and over increased in the 1990s from slightly more than 378,000 in 1990 to almost 576,000 in 2000. However, from 2000–2009, it continually decreased until reaching 300,000, a direct result of the 2008 global crisis on the US economy and the tightening of immigration policy (Gobierno de México 2014). On the other hand, informal sector work has over the last decade averaged at about 60% of the economically active population, considered as employed people who lack access to health care services (Centro de Estudios de las Finanzas Públicas 2014).

These conditions have also affected the quality of formal employment. On the one hand, real wages in the country have deteriorated steadily since 1980 as they lost almost 80% of their purchasing power between 1980 and 2012. On the other hand, precarious employment in the country has expanded, something that ENOE (National Survey on Occupation and Employment) in

Mexico registers as people in critical conditions of employment. The figure reached 12 % of the population in 2013 and includes people with incomes below that of the national minimum wage, working less than 35 hours, in addition to those who work more than 48 hours per week, earning up to two minimum wages (Instituto Nacional de Estadística y Geografía 2014).

Mexico is the most unequal country among the OECD member countries. The average income of the highest earning one-tenth population is 27 times higher than the decile that earns the least in Mexico, while for the OECD overall, that ratio is 10 to 1 (Moreno and Krozer 2015). In this context of economic exclusion and social inequality, crime has expanded significantly in recent years. If the survival alternatives for the population have been strangled in today's economy, crime has become an important field of labour recruitment, especially for the young. This has been further aggravated by the security strategy implemented by the two last governments that has prioritized making war against large criminal organizations, but without any clear intelligence strategy or credible and effective crime prevention strategy.

A report by Global Financial Integrity (2012: 3–16) concludes that these movements, resulting from corruption, bribery, criminal activity and efforts to conceal wealth, accounted for 5.2 % of GDP between 1970 and 2010 (Table 11.1). The annual amount of illicit money leaving Mexico was $3 billion in 1970, increasing to about $10.4 billion in the decade of the 1980s, and on up to $17.4 billion in the 1990s, and by 2000 it reached $49.6 billion per year (Global Financial Integrity 2012: 3–16). According to the organization Mexico United Against Delinquency,[2] crimes considered to have a high social and economic impact in the country (robbery, fraud, homicide, kidnapping and extortion) increased from about 700,000 in 2006 to about 900,000 in 2012, with a specific weight of 53 % on the total number of crimes committed in the country.

The most dramatic aspect of this rise of the criminal economy is that it has been consolidated thanks to enormous social violence. Amnesty International reports that the number of people missing in Mexico since 2006 amounts to 22,610, and almost half of those have disappeared between 2012 and 2014, i.e., during the administration of Enrique Peña Nieto (Amnesty International 2015: 9–12). We will return later to discuss the consequences of this massive increase in crime and social violence for democracy and the rule of law.

The effects of the neoliberal pattern of accumulation in Mexico can be summarized as follows: a) an economy with growth rates lower than the needs of the growing population of working age; b) a structural inability to generate

---

2   *México Unido Contra la Delincuencia.*

TABLE 11.1   *High impact crimes in Mexico, 2006–2013*

Crime	2006	2007	2008	2009	2010	2011	2012	2013
Theft	545,251	610,730	656,877	680,566	738,138	749,414	709,259	511,154
Serious bodily injuries	174,738	189,383	186,585	183,421	176,451	160,995	154,563	114,434
Assassinations	11,806	10,253	13,193	16,117	20,585	22,480	21,728	13,834
Kidnappings	733	438	907	1,162	1,284	1,344	1,317	1,205
Extortions	3,157	3,123	4,875	6,332	6,375	4,404	7,272	6,049
Total high impact	736,685	813,927	862,437	887,598	942,833	938,637	894,139	646,676
Total crimes	1,580,730	1,724,319	1,763,464	1,796,737	1,838,109	1,827,373	1,702,178	1,248,707
Variation %	46.6%	47.2%	48.9%	49.4%	51.3%	51%	53%	52%

SOURCE: MÉXICO UNIDO CONTRA LA DELINCUENCIA

employment; c) a steep and steady decline in real wages; d) a proliferation of precarious employment; e) increasing social inequality and economic exclusion; f) increasing informality and migration as survival strategies; and g) a growth of the criminal economy and criminal activity with an increase in social violence.

How does this state of affairs affect Mexican society in the kind of relationships that develop between the state and civil society? More concretely, in a society like Mexico, are there any real prospects for democracy in the future?

Problems of Democracy in Neoliberalism

A regime is fairly characterized by the type of relations established between the state and society. Such relationships are closely related to the way in which an economy is organized and shapes the struggle for social surplus in the state. So far, this latter element has been characterized very generally by concentrating on its social effects. The focus will now shift to the former.

Up until the 1980s in Mexico, the main legacy in force was that of *Cardenismo*.[3] This was a regime based on a presidency without an effective coun-

---

3   *Cardenismo* refers to the political system attributed to Lázaro Cárdenas (1895–1970), a general in the Mexican Revolution who later became president (1934–1940). Cárdenas "revived pop-

terweight from congress and the judiciary. It was effectively a state-party that annulled pluralism in favour of a strongly hierarchical, corporatist apparatus that extended out from the party. *Cardenismo* unified the vast majority of workers in the countryside and in the urban areas through a strongly hierarchical apparatus that played a mediating role between the state and the popular base, with the electoral system subordinated to the presidential will.

It was in this context that a social pact was established, albeit a limited one, that would prove durable, and in certain reformist periods would allow substantial mobility for its popular base. It was a form of political organization that corresponded to a historical form of economic organization, namely, industrialization with import substitution. A largely successful economic growth pattern was achieved with growth rates exceeding 6% a year; expansive employment rates, even if insufficient; increasing real salaries that accompanied a significant redistribution of wealth; all based on a social policy making apparatus mediated by the corporatist logic, and hierarchically assigned according to the weight each organization held for the economic activities being prioritized for the economy (Enrique de la Garza 2007).

The shift to the neoliberal pattern of accumulation in the context of the debt crisis of the 1980s did not mean a radical transformation of the traditional forms of relations between the state and larger society at the outset. At least in the government of Carlos Salinas, a reformist attempt was made in order to keep the old regime compatible with the transformations in the organization of the economic functions of the state, namely, a social liberalism. This hybrid entered into crisis by 1994 with the armed uprising in the Mexican southeast and political assassinations of that year that would soon combine with the economic crisis of 1995.

---

ulism as a force in national politics by redistributing land to landless peasants under a state-sponsored reincarnation of communal farming known as the *ejido* system. Cárdenas also emphasized nationalism as a force in Mexican politics by expropriating the holdings of foreign oil corporations and creating a new national oil company. Cárdenas' reforms of the late 1930s bolstered the legitimacy of the government while further concentrating power in the president and the Institutional Revolutionary Party (Partido Revolucionario Institucional— PRI), the 'official' party of the Revolution. By the early 1940s, the political processes and institutions that would broadly define Mexican politics for the next forty years were well established: a strong federal government dominated by a civilian president and his loyalists within the ruling party; a symbiotic relationship between the state and the official party; a regular and orderly rotation of power among rival factions within a de facto single-party system; and a highly structured corporatist relationship between the state and government-sponsored constituent groups." "Cardenismo and the Revolution Rekindled, 1934–40" in Merrill and Miró (1996).

Since that time, a process of political transition has been initiated through political electoral reforms as negotiated among the elites of the country's three main political parties. This opened the way for a political alternation with the arrival of an opposition candidate to the presidency in 2000. In this manner, the conditions were created for reversing the order of factors. While in the previous period, there was some inclusion in the economic sphere in the context of an underdeveloped economy, it was effectively in exchange for limitations on political and civil rights and freedoms. In the new period, there was now a political opening that sought to encourage electoral competition and plurality, but it was brought about in exchange for an exclusion of the majority from the benefits of social surplus.

In this context, democracy remains inoperative to the extent that it has diminished its capacity as a legitimizing mechanism of power, indeed, doing so in such a way that the political opening of the previous period has been bogged down and even shows serious signs of varying towards a new form of authoritarianism. In the following section, discussion now turns to analysing some key factors that have democracy "on the ropes" in Mexico, thus undermining the conditions for its viability.

The Crisis of Political Parties

An initial element to be taken into account is the evolution of political parties in this period. Historical studies show a close relationship between the emergence of the modern representative state and the emergence of political parties. In their origins, these organizations were far from being democratic and were built similarly to factions within the parliaments of the first Western democracies until they became institutionalized and competed for power in an organized way. This first period corresponded to a kind of democracy of "notables" that remained confined in amplitude and equality of rights to the feudal owners and bourgeois proprietors. In the countries of the periphery, this type of representation assumed an openly oligarchical form and in Mexico, this period corresponds to the parties during the *Reforma*[4] period up to the *Maximato*.[5]

---

4   The *Reforma* is usually considered to have begun with the overthrow and exile of President Antonio López de Santa Anna in the Revolution of Ayutla in 1854. "A provisional government was installed under Juan Ruiz de Álvarez and the intellectuals of Ayutla; the ensuing period of liberal rule came to be known as The Reform. The Reform was touted as a Mexican version of the French Revolution. Several laws, known collectively as the Reform Laws, abolished the *fueros*, curtailed ecclesiastical property holdings, introduced a civil registry,

A second period at the end of the nineteenth century was inaugurated with the powerful emergence of the worldwide labour movement, resulting in the organization of mass parties from workers' organizations. This was the most fruitful period for representative democracy in terms of the extension of political rights through universal, free and secret suffrage and the right to vote, as well as the extension and recognition of various civil liberties such as freedom of thought, freedom of movement, free press and opinion, freedom to organize and to protest, etc., etc. It was moreover a constant extension of citizenship that in a second phase during the post-WWII war period became expanded to the recognition of a set of social rights through the welfare state (Offe 1994). This involves part of the civil society that organized and through long struggles managed to represent its corporate interests in the state to become a political project with a broad scope. In the developing countries, the emergence of mass parties was mediated by nationalism and corporatism and this was the

---

and prohibited the church from charging exorbitant fees for administering the sacraments. The Reform Laws polarized Mexican society along pro- and anti-clerical lines at a time when delegates were preparing the constitution of 1857, as provided for in the Plan of Ayutla. The new constitution was derived from that of 1824, but it reflected a more liberal vision of society through its incorporation of the Reform Laws. It reaffirmed the abolition of slavery, secularized education, and guaranteed basic civil liberties for all Mexicans." "The Revolution of Ayutla and the Reform Laws" in Merrill and Miró (1996).

5   The *Maximato* (1928–1934) was the period in which Plutarco Elías Calles exercised behind the scenes control over Mexican politics through the actions of three presidents who were essentially his puppets. By 1929 Calles's political machine had found institutional expression as the National Revolutionary Party (*Partido Nacional Revolucionario*—PNR). Unlike previous parties, which existed only in name during electoral campaigns and dissolved immediately thereafter, the PNR was designed to be a permanent organization run exclusively by Calles as *jefe máximo* (supreme leader), through which he acted as de facto president. Henceforth, the "official" party of the revolutionary regime served as the dominant political organization in the country and the primary dispenser of official patronage. The last two years of the *Maximato* under the presidency of Rodríguez witnessed a steady rightward drift of the revolutionary regime. Deciding that the country could not forego agricultural productivity for the sake of equity, Calles ordered a near halt to further land redistribution. Organized labour, which was seen as overly sympathetic to bolshevism and not loyal enough to the PNR, was disavowed and suppressed. By the early 1930s, the government was persecuting the Mexican Communist Party. As the 1934 presidential elections approached, Calles came under increasing pressure from the left wing of the PNR to pursue with more vigor the social welfare provisions of the constitution of 1917. Seeking to avoid a party split, Calles mollified his party's left wing by nominating Lázaro Cárdenas, a popular state governor, to succeed Rodríguez. Calles, although wary of Cárdenas, nevertheless expected the new president to fall into line much as his three predecessors had done. "The *Maximato*" in Merrill and Miró (1996).

case of the PRI in Mexico during the period that began with *Cardenismo* and became extended until the beginning of the 1990s, characterized by a strong vertical organization in its boom period, with selective relations with the state depending on the importance of different workers' organizations.

From the middle of the last century, the political parties became transformed into electoral machines with strong bureaucratic structures (Panebianco 1993). This ended up turning them into what political scientist Otto Kirchheimer called "trapped parties" whose mode of operation is based on obtaining votes beyond its natural electorate in the quest to win elections at the expense of representative integrity and in order to reduce ideological and identity differences. This fundamental element helps explain the recent discrediting of representative democracy under neoliberalism, precisely because the parties stopped serving as mechanisms to transmit the demands of society towards the state. In Mexico, that transition and political alternation brought about an opening of a more pluralistic party system, which, although it transformed the state party into just another competitive party, the new party system acquired the characteristics of the Kirchheimer model.

Richard Katz and Peter Mair (2004) noted the emergence of a new facet of political parties since the mid-1980s, characterized by a situation in which there are no significant differences dividing political parties, no matter how strongly they sometimes compete between each other. It is effectively the emergence of party cartels that enter into a gradual but inexorable withdrawal from the realm of civil society towards the state. In Mexico, this stage was inaugurated in the current six-year period when the *Pacto por México* was signed (2012) and what emerged were parties such as the Green Ecological Party that operate as family businesses, well at the margins of the needs of the population.

The function of operating as an institutionalized mediator in a corporatist model between the workers and the state (as formerly filled by the PRI) was assumed by the cartel parties (the PRI was transformed into one of those) that became clientelist and prebendalist, discontinuous across time as defined by electoral periods, and circumscribed by congressional approval of specific laws. The power of political parties in Mexico now emanates more from the monopoly they exert over politics in general[6] and from representation in particular, and less from popular support (see Figure 11.2).

---

6 While the electoral reform incorporates independent candidates, the regulation of their candidacies is so unequal relative to the candidacies of the political parties as to practically nullify them.

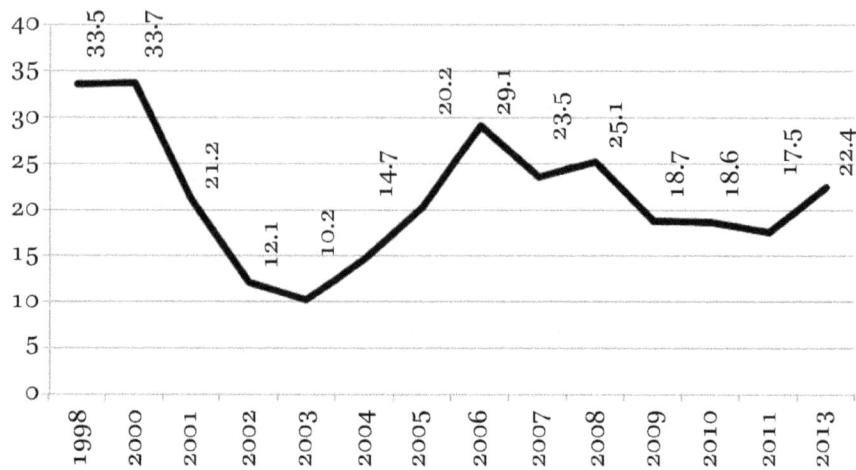

FIGURE 11.2   *Level of confidence in political parties, 1998–2013 (percentage of persons that*
have a lot of / some confidence)
SOURCE: DATA FROM LATINOBARÓMETRO (2015)

There are various processes that have determined the current situation of
the political parties but perhaps the most important would be the individual-
ism that neoliberal ideology has managed to impose in all areas of social life.
This has resulted in general apathy and a growing indifference towards politics
and the common good. This in turn corresponds to the collapse of traditional
forms of organization and representation, such as corporate unions, the social
state, and the developmentalist model of economic growth. On the other hand,
there is the increasing inability of governments to implement autonomous
policies in the context of a colonialism reinforced by foreign governments and
international financial organizations. The parties have confronted the erosion
of their social bases with the security offered by state institutions to politicians willing to reach agreements to share government, programs and voters. In this context, decision-making has migrated to non-majority institutions
as autonomous agencies. With the absence of a real opposition, this form of
democracy does not count for much.

Limited and Unequal Citizenship

A key element of democracy is the establishment of some type of relationship
between the state and citizens where the recognition and enforcement of rights
and obligations for the adult population is broad and egalitarian. It can be said
that the political transition and alternation managed to advance in the design

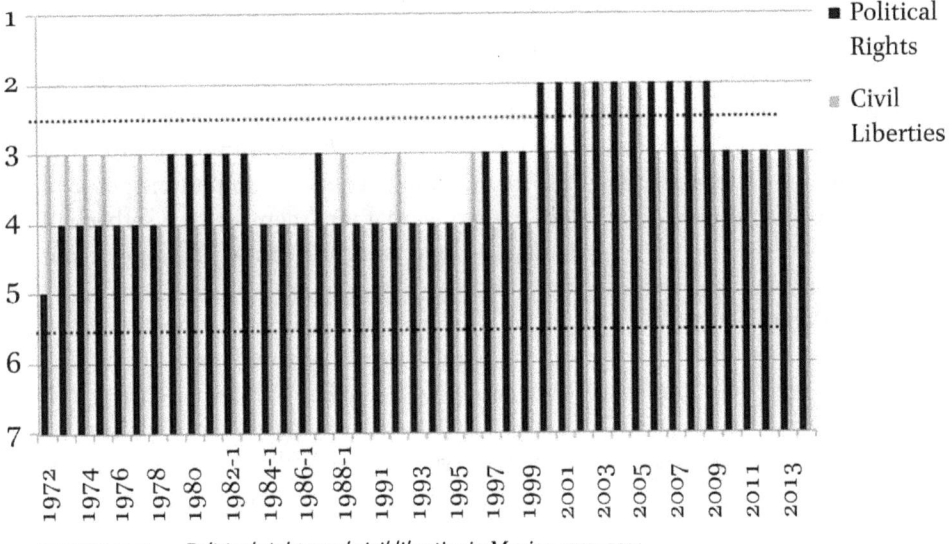

FIGURE 11.3   *Political rights and civil liberties in Mexico, 1972–2014*
SOURCE: FREEDOM HOUSE (2015)

of an institutional and normative framework largely protective of political and civil rights, which had been largely violated under the previous political regime.

This notwithstanding, this legal equality for citizens that is widely recognized has not effectively entered into force. Since 1972, the Freedom House organization has applied a study that registers the level of freedom of countries based on the situation of political rights and civil liberties. The index assigns each country values on a scale from 1 (high) to 7 (low) according to a series of criteria that include compliance with rights such as universal, free and secret suffrage, pluralism, right of minorities to be represented, and a whole set of freedoms including the right of assembly, to protest, to politically organize, protection of property rights, freedom of press and open expression, rule of law and accountability, among others. Based on this index, Freedom House grants the status of a "free" country to those with indices that are in the range of 1.0–2.4; "partially free" to countries with an index between 2.5 and 5.5, and "not free" to those countries that are below 5.5 (Freedom House 2015).

As can be seen in Figure 11.3, Mexico throughout almost the entire history of this organization's rating process has been classified as a "partially free" country, which can be translated as a country with varying levels of authoritarianism. The only period during which the country was classified as a free country was in 2000–2009, precisely during the first two governments of political alternation. Within this period, scores were higher than 2.5 in the areas of civil liberties only for the 2002–2005 period.

The conclusion is obvious. In Mexico, political rights and civil liberties for the population as a whole were present only for a brief period of time, i.e., authoritarianism had prevailed before and after the imposition of the neoliberal pattern of capitalist accumulation. The most important point to note here is that beginning from 2010 forward, there seems to be a setback in what was achieved in the period 2000–2009, since the country has once again been classified as a "partially free" country (Figure 11.3).

To this deficit in the area of political and civil rights must be added the violation of social rights such as the right to work and to a living wage caused by the poor functioning of the economy, the right to decent housing and education given the shrunken budgets for programs in these areas, and the right to health and social protection, with a system selectively aimed at the urban wage-earning population that barely protects just over a third of the working-age population.

Under these conditions, democracy can hardly be consolidated. Short of a broad and egalitarian citizenship, what we have is a political regime in which the type of relations between the state and citizens is defined by extensive areas of social exclusion and violation of the rights of the population where clientelistic relations prevail, limited across time by electoral periods, perpetually at the margins of the law, and therefore, vertical and authoritarian.

Limited Citizen Participation

The participation of the adult population in public affairs within a participatory democracy has a central aspect in the political manoeuvrability of citizens and their ability to influence those who make the political decisions, thereby granting legitimacy to those who govern. According to the study "The Quality of Citizenship in Mexico," conducted by the now extinct Federal Electoral Institute (IFE) in 2013, Mexico registered an abstention rate of more than 40% in the elections of federal deputies between 1982 and the time of their study, with 1982 and 1994 being the electoral years with the lowest abstention (Instituto Federal Electoral 2013). In the first case, this can be explained by the effects of the 1977 reform. In the latter case, the so-called "fear vote" in favour of the PRI was observable in the context of intense political violence across the country. The 1988 and 2003 elections, on the other hand, showed an abstention rate of over 50%. The former can be explained due to the total lack of transparency

TABLE 11.2   *Electoral participation*

Country	% of registered voters that vote	Year of election (congress or parliament)
Chile	49.35	2013
Germany	71.55	2013
Spain	68.94	2011
Argentina	79.39	2011
Canada	61.49	2011
United Kingdom	65.77	2010
United States	67.95	2012
France	55.40	2012
Mexico	62.08	2012

SOURCE: INSTITUTO FEDERAL ELECTORAL (2013)

with which the votes were counted in the election of that year, while in 2003, the abstention can be explained because it was an intermediate election.

Subtracting out these peak values, there is a significant average abstention rate, although not entirely exceptional among those recorded in the same years in consolidated democracies such as France, United States, United Kingdom and Canada. (see Table 11.2). In the case of democracies with presidentialist systems of government and a uninominal electoral system of representation, the effects of this abstention have had more important and lasting effects since the executive governments govern with the support of a minority proportion of the electorate as has occurred in Mexico over recent decades. On the other hand, non-electoral citizen participation has also remained at a low level in Mexico as evidenced by the IFE study results that are summarized in Table 11.3.

It is worth noting that non-electoral participation through political parties accounts for 20% while those activities that do not require such a great effort, such as talking about politics, reading or sharing information in social networks, accounts for 49%. In this sense, the activities that could have a more direct effect on politics through non-formal channels such as strikes or social protest have only marginal participation. This is due in large part to the inhibiting effect of the criminalization of social protest in Mexico over recent years and the increasingly secondary role played by trade union organizations. The regime has abandoned collective forms of political participation in favour of individual forms while neoliberalism has done its own part pro-

TABLE 11.3   *Types of non-electoral political participation in which Mexicans engage*

Type of participation	%
Chat with other people about political matters.	39
Participated in local or municipal meetings of political advocacy.	12
Collaboration with political parties prior to or during electoral campaigns.	11
Concerted attempts to convince friends to vote for a specific candidate.	11
Read or shared political information by a social media network such as Twitter or Facebook	10
Signed petitions or documents designed to express a protest.	9
Participated in demonstrations or public protests.	6
Occupied or blockaded public places or installations.	3
Participated in a strike.	2

SOURCE: INSTITUTO FEDERAL ELECTORAL (2013)

moting the internalization of the idea that human beings are self-interested, profit-maximizing beings who pursue their own good without regard to the well-being of the larger society.

Rule of Law

This has had unexpected consequences in the structuring of the social fabric of Mexico. The levels of violence have reached levels never before seen, shadowed by alarming levels of impunity that have engendered the historical corruption that prevails in the country's institutions. Earlier, we had pointed to the rule of law as being a fundamental element of representative democracy, i.e., the protection of law for the citizen against the arbitrariness of the state. It should be emphasized that the law in itself has never guaranteed political equality, much less economic equality to citizens of a territory. More typically, it codifies the social inequalities that prevail in a society. In this sense, not every society with the rule of law can automatically be categorized as a democracy. Democracy cannot be simply summed up to a legal equality of citizens before the law, but to the fact that the process of government effectively allows it to reduce the structures and forms of inequality and discrimination, beginning with an opening in the processes of decision making (Tapia 2011: 170). Given this fact, it is also true that democracy can hardly exist without

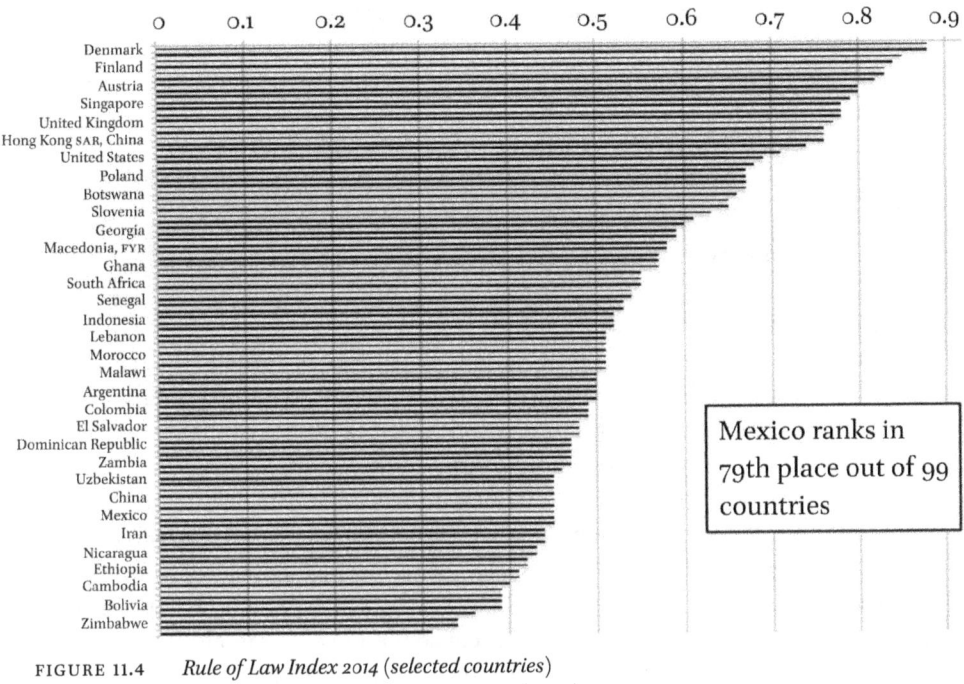

FIGURE 11.4 *Rule of Law Index 2014 (selected countries)*
SOURCE: WORLD JUSTICE PROJECT (2015)

the rule of law. For a regime to be authoritarian, it must effectively operate above the law. This is precisely what Mexico has experienced in recent years and that is why it is important to include it in the diagnosis of Mexican democracy.

The Rule of Law Index generated by the World Justice Project is based on the measurement of nine factors, including: counterweights to government power; absence of corruption; fundamental rights; order; security; regulatory compliance; civil justice; criminal justice; and informal justice. These factors are disaggregated into 42 distinct variables in the index that is measured on a scale of 0 to 1, where the lowest level in meeting the criteria is 0 and the highest is 1 (World Justice Project 2015). Mexico ranks 79th out of the 99 countries that make up the study with an index of .48. This is very far from Denmark, Norway, Sweden, Finland and the Netherlands that lead the ranking with indexes above .8. Below are Tanzania, Zambia, Kazakhstan, Uzbekistan, Egypt, Moldova, China, Ecuador and Kyrgyzstan while just above are Russia, Madagascar, Iran and Guatemala, among others. (See Figure 11.4)

This is another of the most sensitive pending issues of democracy in Mexico since corruption and impunity, which are the main problems evidenced by the rule of law in Mexico and recorded in this study, actually constitute state crime that generates violence toward the population from the state and is further elevated on account of existing in a highly unequal society. The main danger today for democracy is that relations between the state and society are increasingly mediated by a spike in crime that engenders violence.

Conclusion

Representative democracy in Mexico faces enormous challenges that are difficult to resolve. The form of economic organization developed over recent years has imposed enormous barriers to the establishment of a broader, more equal, and more protected type of state-citizen relationship that would allow for actively effective participation in public affairs.

The neoliberal pattern of accumulation has been consolidated in Mexico at the cost of generating high rates of exploitation of salaried workers. This includes the displacement and social exclusion of large social sectors that have been marginalized and relegated to survival at the side-line of the formal processes of capital appreciation. The consolidation of criminality as a thriving branch of the economy has become embedded across all branches of social life. State institutions, for their part, operate behind the backs of citizens in a system riddled by corruption and characterized by opacity.

The counterpart of these processes has been the emergence of a set of social uncertainties: unemployment and precariousness in working conditions; high levels of informal economy; high levels of illegal economic activities; and heightened uncertainty over life, property and freedom in the face of pervasive criminality and the absence of the rule of law. This set of uncertainties has operated as an effective mechanism of domination over large swaths of Mexican society in recent years, creating a strong deterrent effect on the participation, organization and autonomous action of citizens to influence public affairs— whether through formal channels such as elections, or by informal means such as social movements—thereby contributing to reproducing the existing social order.

Governments resulting from the elections of recent years have operated behind the backs or in outright opposition to the interests of the majority of the population. Representative democracy in this context has totally lost its content. There is no accountability but only opacity as important decisions have been relegated to bureaucratic power spaces that are not democratically elected, and worse yet, where there are powerful foreign states and international agencies that strongly influence state decision making on behalf of their own financial investments. The *Pacto por México* and the introduction of reforms without consultation are an example of this syndrome to which a widespread, violent criminality has become added into the political mix as evidenced by the horrendous events of Tlataya, Ayotzinapa[7] and San Quintín.[8]

There is a clear disconnect between citizens and the government in Mexico. Mexican citizens do not feel represented, nor do they have confidence in the most visible institutions of the state. Only 19% place their trust in a political party, 17% trust their deputies, and 32% trust the police. In addition, 65% do not have access to any intermediary that allows them to avail themselves to justice, government, or community political resources. Democracy requires

---

7   "On June 30, 2014 twenty-one men and one woman were executed by soldiers in Tlataya, a rural area in Mexico State. The local governor maintained that the army had, in 'legitimate self-defense, taken down the criminals.' One witness, whose daughter was among the dead, claimed that soldiers had in fact lined up the twenty-two people before executing them one by one. The eyewitness said she told the soldiers not to do it, not to kill those being interrogated. Their response, she reported, was that 'these dogs don't deserve to live.' The cover-up that ensued involved bureaucrats from various levels of government. It was only because of reporting by *Esquire* magazine and the work of local journalists in Mexico that the truth about what happened in Tlatlaya came out. Eight soldiers are suspected of being directly involved in the killings. Seven have been charged, three of them for murder. The massacre in Tlatlaya was quickly overshadowed by another perpetrated by police and gunmen in Iguala, Guerrero. On the night of September 26, 2014, six people were killed, three of them students at a nearby teacher-training college. One of the young men who was killed had his face pulled off and pulled down around his neck. Others were denied medical treatment. By the next day, forty-three more students from the Raúl Isidro Burgos Rural Teachers' College of Ayotzinapa were missing. All of the students were last seen as they were in the process of being arrested by municipal police, allegedly for participating in taking over buses to use for transportation to a march in Mexico City. The police handed the students off to a local paramilitary group, which the media dubbed Guerreros Unidos (United Warriors)." The case remains unsolved. See Paley (2015).

8   The brutality in San Quintín came during a wave of violent repression against ongoing rural protests for a living wage and better working conditions for farm day labourers, leaving three dead. Police forces used rubber bullets, tear gas, and gunfire against striking farmworkers that began early the day after a private ranch owner called the police about a crowd of people who had gathered to urge their fellow workers to continue the strike. The harsh police presence, which included an unannounced raid on workers' homes in the community, triggered further protests and clashes throughout the day. Over seventy were injured, with seven hospitalized in critical condition. Three others were killed when police opened fire, according to farmworkers. Rural organizations in San Quintín are holding state government officials, agribusiness companies operating in the area, and the police officers who carried out the repression responsible for the recent violence and injuries. See TeleSur (2015).

the support and agreement of citizens, and therefore a climate of social trust and cooperation. However, 27% of Mexicans said that they did not trust the majority of people. It is no coincidence that most Mexicans have an idea of democracy as a type of regime in which "many play, but few win" (Instituto Federal Electoral 2013: 198–199). What kind of country can you build with that? What is to be done?

It is difficult to see how a democracy can be constructed on the basis of capitalism, particularly under the current pattern of accumulation. Western democracies are going through a crisis at different levels of intensity and there is a contradictory logic between political and juridical equality that promotes representative democracy and the inequality that has been generated by neoliberal globalization. These crises represent two dimensions of the same problem: economic organization and political regulation.

It seems that the answers are not where we always believed they were to be found. Modern history, and Mexico is no exception, has shown us that the processes of democratization were driven in large part by popular struggles. The journey traversed by Greece, Spain, Italy as well as in North Africa and South America, questioning capitalism, is an irreversible march and is forcing the emergence of another type of democracy that challenges the presently existing order that has become stagnant. Greater participation is what is needed but it must be based upon new foundations and a new social pact.

References

Abahaj, Safia. 2015. "El eterno Xi Jinping", nuevatribuna.es—Sección Internacional, 11 April. [Accessed 1 July 2015]. Available at: http://www.nuevatribuna.es/articulo/mundo/eterno-xi-jinping/20150411184431114630.html

Acosta Reveles, Irma Lorena. 2006. "Balance of Agro-export Models in Latin America at the Beginning of the XXI Century." *World Agricultural Magazine, Journal of Rural Studies*. 13 (July–December: 1–25). [Accessed 3 March 2015]. Available at: http://www.scielo.org.ar/scielo.php?script=sci_arttext&pid=S1515-59942006000200001

Acosta Reveles, Irma Lorena. 2010. *Latin America: Capital, Labor and Agriculture on the Threshold of the Third Millennium*. Mexico: Miguel Angel Porrua-UAZ.

Acosta Reveles, Irma Lorena. 2013. "The scientific and technological factor in the consolidation of regional agrarian capitalism." *Journal of Rural Development*. 71 (July–December: 15–35). [Accessed 5 February 2015]. Available at: http://dx.doi.org/10.1144/6264

Agencia de Noticas Xinhua. 2010. "Firmas de capital extranjero fabrican casi el 70% de maquinaria y productos electrónicos exportados por China," Agencia de Noticias Xinhua-Spanish.News, February 16. [Accessed 24 August 2015]. Available at: http://spanish.news.cn/tec/2010-02/16/c_13177186.htm

AGROBIO (Association of Agricultural Plant Biotechnology). 2014. "Global Status of Biotech Crops in 2014." AGROBIO. [Accessed 22 April 2015]. Available at: http://isaaa.org/resources/publications/briefs/49/executivesummary/default.asp

Aguirre, Pedro. 2009. Sistemas políticos y electorales contemporáneos. Corea del Sur. Vol. 25, Mexico: Instituto Federal Electoral.

Alcántara, José. 2010. *La neutralidad de la red*. Bilbao: Biblioteca de las Indias.

Alzugaray, Carlos. 2004. "De Bush a Bush: balance y perspectivas de la política externa de los Estados Unidos hacia Cuba y el Gran Caribe." Pp. 201–244, in José María Gómez, ed., *América Latina y el (des) orden global neoliberal Hegemonía, contrahegemonía, perspectivas*. Buenos Aires, Argentina: CLACSO.

Amnesty International. 2014. "Libertad religiosa", China—Sección Española [Accessed 6 June 2014]. Available at: https://www.amnesty.org/es/countries/asia-and-the-pacific/china/report-china/2015. China: No End in Sight—Torture and Forced Confessions in China (Executive Summary). [Accessed 6 June 2016]. Available at: https://www.amnesty.org/en/documents/asa17/2731/2015/en/

Amnesty International. 2015. *Información para el Comité contra las Desapariciones forzadas de la ONU*. Madrid: Amnesty International Publications.

APF. 2013. "Aumentan emisiones de dióxido de carbono de carbono (CO2) a causa del carbón," El Universal, 19 November. [Accessed 1 February 2014]. Available at: http://www.eluniversal.com.co/ambiente/aumentan-emisiones-de-dioxido-de-carbono-co2-causa-del-carbon-142370

Bambirra, Vania. 1978. *Teoría de la dependencia: una anticrítica.* México: ERA.

Barsky, Osvaldo. 1980. *Proyecto cooperativo de investigación sobre tecnología agropecuaria en América Latina.* Quito, Ecuador: FLACSO.

Bauman, Zygmunt. 2006. *Amor líquido.* Argentina: Fondo de Cultura Económica.

Bavoleo, Bárbara and Ladevito, Paula. 2009. "Sociedad Civil y Proceso de Democratización En Corea del Sur," in *Estudios Internacionales,* No. 164: 79–93. [Accessed 18 February 2014]. Available at: http://www.revistaei.uchile.cl/

Bojórquez, Nelia. 2005. "Ciudadanía," in *Derechos de la infancia en riesgo,* Yolanda Corona Caraveo and Norma del Río Lugo (eds.), UAM, Universidad de Valencia. [Accessed 18 February 2014]. Available at: http://www.uam.mx/cdi/derinfancia/5nelia.pdf

Borón, Atilio. 2000. *Tras el Búho de Minerva. Mercado contra democracia en el capitalismo de fin de siglo.* Buenos Aires, Argentina: Fondo de Cultura Económica. http://bibliotecavirtual.clacso.org.ar/ar/libros/buho/biblio.rtf

Brynholfsson, Erik and McFee, Andrew. 2014. *The Second Machine Age: Work, Progress and Prosperity of Brilliant Technologies.* New York and London. W.W. Norton & Company, Inc., https://tanguduavinash.files.wordpress.com/2014/02/the-second-machine-age-erik-brynjolfsson2.pdf [Accessed 10 May 2015].

Bureau of Labor Statistics. Data Tools. 2014. Department of Labor, Washington, DC. https://www.bls.gov/data/ [Accessed 28 April 2015].

Bureau of Labor Statistics. Employment Projections 2012–2022. 2013. Department of Labor, Washington, DC. http://www.bls.gov/opub/mlr/2013/article/occupational-employment-projections-to-2022.htm [Accessed 23 March 2015].

Bustelo, Pablo. 1991. "La expansión de las grandes empresas de corea del sur, (Chaebol), un ejemplo de estrategia corporativa." In *Cuadernos de Estudios Empresariales.* Núm. 1, No. 8 (Jan.–Feb.).

Cabrero, Enrique, Sergio Cárdenas, David Arellano, and Edgar Ramírez. 2011. "La vinculación entre la universidad y la industria en México. Una revisión a los hallazgos de la Encuesta Nacional de Vinculación," *Perfiles Educativos:* 186–199.

Calva, José Luis. 2005. "México: la estrategia económica 2001–2006. Promesas, resultados y perspectivas," in *Problemas del Desarrollo,* No. 143 (October–November): 59–89. [Accessed 10 March 2012]. Available at: http://www.ejournal.unam.mx/pde/pde143/PDE14303.pdf

Calvo Ospina, Hernando. 2010. *El Equipo de Choque de la CIA.* Barcelona, España: El Viejo Topo.

Campos Ríos, Guillermo and Germán Sánchez Daza. 2005. "La vinculación universitaria: ese oscuro objeto de deseo," *Revista Electrónica de Investigación Educativa* 7(2): 1–13.

Campos Ríos, Guillermo and Germán Sánchez Daza. 2006. "La vinculación universitaria y sus interpretaciones", *Ingenierías* IX, No. 30: 18–25.

Cardoso, Fernando H. 1972. "Dependency and Development in Latin America." *New Left Review*, 74: 83–95.

Carr, Nicholas. 2011. *Superficiales: ¿qué le está haciendo internet a nuestras mentes?* Madrid: Editorial Taurus.

Casas, Rosalba and Matilde Luna. 1994. "Condicionantes políticos de la nueva relación entre universidad e industria," pp. 1–17 in *Universidad y vinculación: nuevos retos y viejos problemas*, Miguel Angel Campos and Leonel Corona, eds. Mexico: UNAM.

Castaños-Lomnitz, Heriberta. 1997. "Reluctant partners in Modernization: The National Autonomous University of Mexico and its Links with Industry," in *Higher Education*, 33(4): 363–379.

Castells, Manuel. 2006. *The Theory of the Network Society*. Great Britain: MPG Books, 2006.

CAWMA (Comprehensive Assessment of Water Management in Agriculture). 2007. *Water for Food, Water for Life: A Comprehensive Assessment of Water Management in Agriculture*. London: Earthscan.

Centro de Estudios de las Finanzas Públicas. 2015. "Indicadores sobre Seguridad Social en México". Centro de Estudios de las Finanzas Públicas. [Accessed 25 June 2015]. Available at: http://datos.imss.gob.mx/dataset?query=

Chang Castillo, Helene Giselle. 2010. "El modelo de la triple hélice como un medio para la vinculación entre la Universidad y Empresa," in *Revista Nacional de Administración*, núm. 1 (Jan.–June): 85–94.

Chomsky, Noam. 1997. *Secretos, mentiras y democracia. Entrevista por David Barsamian*. México: Siglo XXI.

CHRD. 2014. "Prisoner of conscience—Dong Rubin." Portraits of defenders, Chinese Human Rights Defenders (CHRD). [Accessed 22 November 2014]. http://chrdnet .com/2014/08/prisoner-of-conscience-dong-rubin

CIA (US Central Intelligence Agency). 2014. "Population Below Poverty Line", *CIA World Factbook 2013–14*. Washington, D.C.: United States Central Intelligence Agency. Available in: https://www.cia.gov/library/publications/the-world-factbook/fields/2046 .html

Cochrane, Joe. 2014. "Embrace of Atheism Put an Indonesian in Prison." *New York Times*, Asia Pacific, May 3. Available at: http://www.nytimes.com/2014/05/04/world/asia/ indonesian-who-embraced-atheism-landed-in-prison.html

Colburn, Forrest D. 2002. *Latin America at the End of Politics*. Princeton, NJ: Princeton University Press.

Comisión Económica para América Latina. CEPALSTAT. "Bases de Datos y publicaciones estadísticas". Comisión Económica para América Latina. http://estadisticas .cepal.org/cepalstat/WEB_CEPALSTAT/Portada.asp (Fecha de consulta: 3 de junio de 2015).

Contreras, Sergio. 2013. "Internet: red de mentiras." *Revista Etcétera* 155 (October): 65–

72. Available at: http://www.etcetera.com.mx/articulo/Internet%3A+red+de+
mentiras/22043

Cores, Hugo. 2007. "Former Guerillas in Power: Advances, Setbacks and Contradictions
in the Uruguayan Frente Amplio." Pp. 221–234, in R.A. Dello Buono and José Bell Lara,
eds., *Imperialism, Neoliberalism and Social Struggles in Latin America*. Leiden, Países
Bajos: Brill Academic Press.

Corporación Latinobarómetro. "Banco de datos". Corporación Latinobarómetro.
[Accessed 26 June 2015]. Available at: http://www.latinobarometro.org/latOnline
.jsp

CRI (China Radio International). 2013. "The State Council of China", China ABC.
[Accessed 1 June 2014]. Available at: http://english.cri.cn/7586/2013/10/30/
3441s795053.htm

Cueva, Agustín. 1974. "Problemas y perspectivas de la teoría de la dependencia," pp. 55–
77 in *Historia y Sociedad* 3, México.

Cumings, Bruce. 1997. *Korea's Place in the Sun: A Modern History*. New York: W.W. Norton
& Company.

Da Silva, Carlos, et al. 2013. *Agro-industries for Development*. Rome: FAO. [Accessed
22 May 2015]. Available at: http://www.fao.org/docrep/017/i3125s/i3125s00.pdf

Dahl, Robert. 1971. *Polyarchy: Participation and Opposition*. New Haven, CT: Yale University Press.

Daly, H.E., J.B. Cobb, and C.W. Cobb. 1994. *For the Common Good: Redirecting the
Economy toward Community, the Environment, and a Sustainable Future*. Boston:
Beacon Press.

De Kerckhove, Derrick. 1997. *Connected Intelligence: The Arrival of the Web Society*.
Toronto: Somerville House Books.

de la Barra, Ximena. 2007. "Social and Financial Debt: The Dual Debt of Neoliberalism." Pp. 37–83, in R.A. Dello Buono and José Bell Lara, eds., *Imperialism, Neoliberalism and Social Struggles in Latin America*. Leiden, Netherlands: Brill Academic
Press.

de la Barra, Ximena. 2009. "América Latina Solidaria Ofrece Respuestas a la Crisis."
CENDA. Available at: http://www.rebelion.org/docs/91326.pdf

de la Barra, Ximena and Dello Buono, R.A. 2009. *Latin America After the Neoliberal
Debacle*. Lanham, MD: Rowman and Littlefield.

de la Rua, Ainhoa de Federico. 2010. "La perspectiva del interaccionismo estructural
para el análisis de redes sociales," in *Elementos para el trabajo en red Apuntes desde
el análisis de redes sociales* [digital], J.L. & Maya Jariego Molina, I. (eds.) Barcelona,
Spain: REDES, Revista Hispana para el Análisis de Redes Sociales.

Dedrick, Jason, Kenneth L. Kraemer and Greg Linden. 2010. "Who Profits from Innovation in Global Value Chains? A Study of the iPod and Notebook PCs," *Industrial and
Corporate Change*, Oxford University Press, 19 (1), February.

Dello Buono, R.A. 1995. "An Introduction to the Cuban Special Period." Pp. 1–11, in *CartaCuba: Interdisciplinary Reflections on Development and Society*. Havana, Cuba: FLACSO.

Dello Buono, R.A. 2007. "The Changing Face of Latin America's Political Parties." Pp. 277–299, in R.A. Dello Buono and José Bell Lara, eds., *Imperialism, Neoliberalism and Social Struggles in Latin America*. Leiden, The Netherlands: Brill Academic Press.

Dello Buono, R.A. and Amalia Chamorro. 1990. "The Political Economy of the Sandinista Electoral Defeat." *Critical Sociology*, 17, No. 2: 93–101.

Desai, Meghnad, Amartya Sen and Julio Boltvinik. 1998. Índice De Progreso Social. Una Propuesta, Mexico: UNAM/Centro de Investigaciones Interdisciplinarias en Ciencias y Humanidades.

Diario del Pueblo. 2000. "Consejo de Estado", Sección Guía de China-Órganos estatales. [Accessed 1 June 2014]. Available at: http://spanish.peopledaily.com.cn/spanish/articulos/org/O102.html

Díaz, Mariela P. 2005. "La lucha por la democracia en Corea del Sur", III Jornadas de Jóvenes Investigadores, November. Instituto de Investigaciones Gino Germani, Facultad de Ciencias Sociales, University of Buenos Aires. [Accessed 29 February 2014]. Available at: http://webiigg.sociales.uba.ar/iigg/jovenes_investigadores/3JornadasJovenes/Templates/Eje%20Poder%20y%20Dominacion/Diaz%20Mariela%20-%20Poder.pdf

Dos Santos, Theotonio. 1970. "The Structure of Dependence." *American Economic Review*, 60, No. 2 (May 1970): 231–236.

EFE. 2013. "China aplica restricción a la prensa," CNN Expansión—Sección Economía, 17 April [Accessed 6 May 2014]. Available at: http://www.cnnexpansion.com/economia-insolita/2013/04/17/china-regula-citas-en-medios-locales

EFE. 2014a. "China censura de internet a The Big Bang Theory," El Universal—Sección Espectáculos, 28 April. [Accessed 7 May 2014]. Available at: http://www.eluniversal.com.mx/espectaculos/2014/china-censura-de-internet-a-the-big-bang-theory-1006456.html

EFE. 2014b. "El desempleo de China se mantiene en el 4.1 en 2013," Expansión.com, 24 January. [Accessed 2 June 2014]. Available at: http://www.expansion.com/2014/01/24/economia/1390546342.html

Enrique de la Garza. 2007. "Democracia, representatividad y legitimidad syndical," pp. 9–20, in *Democracia y cambio sindical en México*, Enrique de la Garza, ed., Mexico City: Plaza y Valdés.

Epoch Times. 2013. *The Epoch Times*. [Accessed 2 June 2014.] Available at:

Escalante, Fernando. 2015. "Los años setenta. Breve historia del neoliberalismo," in *Nexos*, 1 May. [Accessed 2 June 2015]. Available at: http://www.nexos.com.mx/?p=24790

Etzkowitz, Henry. 1993. *Academic-Industry relations: A new mode of production?* México: CIT-UNAM.

Etzkowitz, Henry. 2008. *The Triple Helix: University-Industry-Government Innovation in Action.* Routledge.

Etzkowitz, Henry and Loet Leydesdorff. 1997. *Universities and the Global Knowledge Economy: A Triple Helix of University-Industry-Government Relations.* Continuum International Publishing Group.

Feenberg, Andrew. 2010. *A teoria crítica de Andrew Feenberg: racionalização democrática, poder e tecnologia.* Brasília: Observatório do Movimento pela Tecnologia Social na América Latina / CDS / UnB / Capes. SérieCadernos—PrimeiraVersão / construção social da tecnologia / número 3.

Ferrater Mora, José. 1964. *Diccionario De Filosofía* (Tomo I, a-K). 5a. ed. Montecasino. Buenos Aires, Argentina: Editorial Sudamericana.

Ferrer-i-Carbonell, Ada. 2011. "Happiness Economics." Els Opuscles del CREI.

Figueroa Delgado, Silvana Andrea. 2008. "América Latina, otra ruta: el crecimiento desde dentro," *Observatorio de la Economía Latinoamericana*, Universidad de Málaga, No. 106, December. [Accessed 25 September 2014.] Available at: http://www .eumed.net/cursecon/ecolat/la/08/safd.htm

Figueroa Delgado, Silvana Andrea. 2012. "Un ambiente para el desarrollo: el caso de Corea del Sur," in América Latina Globalidad e Integración, ed. by Colomer V. Antonio, pp. 647–653, Madrid, Ediciones del Orto.

Figueroa Sepúlveda, Víctor Manuel. 1996. *Reinterpretando el subdesarrollo: trabajo general, clase y fuerza productiva en América Latina.* México, D.F.: Siglo XXI.

Figueroa Sepúlveda, Víctor Manuel. 2001. "América Latina: el nuevo patrón de colonialismo industrial," *Problemas del Desarrollo* 32, No. 126: 9–33.

Figueroa Sepúlveda, Víctor Manuel. 2015. "Is there any Future for Democracy?" *Critical Sociology* 41(3): 413–422.

Frank, Andre Gunder. 1966. "The Development of Underdevelopment." *Monthly Review,* 18.

Freedom House. 2015. "Freedom in the Word. Individual country ratings and status, FIW 1973–2015 (EXCEL)." Freedom House. https://freedomhouse.org/report/ freedom-world/freedom-world-2015#.VbbdynjldES [Accessed 12 June 2015].

Frey, Carl and Osborne, Michael. 2013. *The Future of Employment: How Susceptible are Jobs to Computerization?* http://www.futuretech.ox.ac.uk/www.futuretech.ox.ac.uk/ future-employment-how-susceptible-are-jobs-computerisation-oms-working -paper-dr-carl-benedikt-frey-m.html [Accessed 3 March 2015].

Fronesis, 2010. "II Pronunciamiento Latinoamericano por una Educación para Todos." *Fronesis,* Septiembre. http://otraeducacion.blogspot.com/p/pronunciamiento -latinoamericano-por-una.html

Fuentes, Claudio, Andrés Villar and Marcela Ríos. 2007. *Dinero y Política: Contribuciones al Debate sobre Financiamiento Electoral.* Santiago de Chile: FLACSO.

FUHEM Ecosocial. 2013: *Land Grabbing Report*. Madrid: FUHEM-Ecosocial.

Fukuyama, Francis. 2003. *Nosso futuro pós-humano—Consequências da revolução da biotecnologia*. Rio de Janeiro: Rocco.

Furtado, Celso. 1974. *Teoría y política del desarrollo económico*. México, D.F.: Siglo XXI.

García Echevarría, Santiago. 1993. *Teoría económica de la empresa*. Madrid, España: Ediciones Díaz de Santos.

García Canclini, Néstor. 1995. *Consumidores y ciudadanos: Conflictos multiculturales de la globalización*. México, D.F.: Editorial Grijalbo.

García García, José Odon. 2012. "Agricultural Activity in Mexico and the World within the Global Food System: Between Agribusiness and Food Supply." *Inceptum*, 13 (July–December: 395–420). [Accessed 5 May 2015]. Available at: http://www.inceptum.umich.mx/index.php/inceptum/article/view/217/197

Garrafa, Volnei and Berlinguer, Giovanni. 1996. *O Mercado Humano—Estudo bioético da compra e venda de partes do corpo*. Brasília: Editora UNB.

Garretón, Roberto. 2011. "Chile: Perpetual Transition under the Shadow of Pinochet," pp. 73–92 in de la Barra, Ximena, *Neoliberalism's Fractured Showcase: Another Chile is Possible*. Leiden, The Netherlands: Brill Academic Press.

Garrido Colmenero, Alberto. 2009. "Water as a Scarce Resource: Defining the Ownership of Water in Consideration of Global, National and Regional Aspects." *Economic Mediterranean* 15: 143–161. [Accessed 4 April 2015]. Available at: http://www.publicacionescajamar.es/pdf/publicaciones-periodicas/mediterraneo-economico/15/15-258.pdf

Giné, Daví Jaume. 2010. "Economía de Corea del Sur en 2009 y perspectivas para 2010." [Accessed 14 March 2014]. Available at: http://www.igadi.org/artigos/2010/jgd_la_economia_de_corea_del_sur_en_2009

Girola, Lidia. 2005. *Anomia e individualismo*. Mexico: Anthropos.

Gliessman, Stephen. 2002. *Agroecology: Ecological Processes in Sustainable Agriculture*. Costa Rica: AGRUCO-CATIE.

Global Financial Integrity. 2012. *México: Flujos Financieros ilícitos, desequilibrios macroeconómicos y economía sumergida*. Washington, D.C.: Ford Foundation.

Gobierno de México. 2014. "Mexican Migration to the United States." Consejo Nacional de Población. [Accessed 15 June 2015]. Available at: http://www.omi.gob.mx/es/OMI/Documentos_de_interes

Gómez Chiñas, Carlos. 2003. "Comercio Exterior y Desarrollo económico. El caso de Corea del Sur." in *Análisis Económico*, Vol. XVIII, No. 37: 141–155.

González Casanova, Pablo. 1992. "La crisis del Estado y la lucha por la democracia en América Latina," pp. 17–38 in *La democracia en América Latina: actualidad y perspectivas*, eds. Pablo González Casanova & Marcos Roitman Rosenmann. Madrid: Editorial Complutense.

González, Susana. 2012. "Vinculadas con universidades, sólo 14 de 100 empresas: SEP,"

La Jornada, 14 January 2012. [Accessed 11 February 2014]. Available at: http://www
.jornada.unam.mx/2012/01/14/economia/024n1eco

Gordon, Colin. 2015. "Wolves of Wall Street: Financialization and American Inequality,"
Dissent, April 17. http://www.dissentmagazine.org/online_articles/wolves-of-wall
-street-financialization-and-american-inequality [Accessed 25 April 2015].

Greenwald, Glen. 2014. *Snowden: sin un lugar donde esconderse*. Nueva York: Metropolitan Books.

Grou, Pierre. 1988. *L'emergence des géants du tiers monde. Contróle et stratégies des
firmes et banques du tiers monde*. Paris: Ed. Publisud.

Guang, Lu. 2009. *Pollution in China*—"Cancer Villages in China" (Film Documentary).
[Accessed 1 June 2014]. Available at: http://www.youtube.com/watch?v=
XHM3soyH7qw

Guimaraes, Lytton. 2010. "El modelo Coreano de Desarrollo y su transferabilidad," in
Centro Estudios Coreanos in *Argentina Newsletter*. [Accessed 4 May 2014]. Available
at: http://www.uba.ar/ceca/newsletters/agosto_10/nl-nota1.php

Habermas, Jürgen. 2004. *El futuro de la naturaleza humana. ¿Hacia una eugenesia
liberal?* Barcelona: Paidós.

Hair, Joseph F. Jr., Ralph E. Anderson, Ronald L. Tatham, and William C. Black. 1999.
Análisis Multivariante. Madrid, España: Prentice Hall/Pearson Educación, S.A.

Han, Sang-jin. 1997. "El fortalecimiento de los sectores populares medios y su futuro,"
in SILBERT, Jaime (ed.), *La República de Corea Hoy: Economía, Sociedad, Relaciones
Internacionales*. Córdoba: Ed. Comunicarte, pp. 15–50.

Han, Sang-jin. 1998. "De la burocracia autoritaria a la sociedad civil: las lecciones
de la experiencia coreana," in Silbert/Santarrosa (comp.) *Desarrollo económico y
democratización en Corea del Sur*.

Harvey, David. 2007. *El nuevo imperialismo. Acumulación por desposesión*. España: Akal.

Hills, P., and M. Argyle. 2002. "The Oxford Happiness Questionnaire: A Compact Scale
for the Measurement of Psychological Well-Being," *Personality and Individual Differences* 33: 1073–1082.

Horkheimer, Max. 1975. *Textos Escolhidos: Filosofia e Teoria Crítica. Coleção Os Pensadores*, volume XLVIII. São Paulo: Abril Cultural. Available at: http://www
.lagranepoca.com/archivo/category/free-tagging/expropiaciones-en-china.html

IFPRI (International Food Policy Research Institute). 2006. *Bioenergy and Agriculture:
Promises and Challenges*. Spain: IFPRI—CGIAR. [Accessed 7 June 2015]. Available at:
http://www.fao.org/fileadmin/user_upload/AGRO_Noticias/docs/bioenergia%20y
%20agricultura.pdf

Ilbo, Chosun. 2004. "A Wave of Impeachment-Related Rallies Sweeps the Nation," in
Digital, Seoul.

INEGI. 2012a. "Bienestar Subjetivo. Microdatos." National Institute of Statistics and
Geographical Data-INEGI. Available at: http://www3.inegi.org.mx/sistemas/
microdatos/default_BN.aspx

INEGI. 2012b. "Presenta El INEGI Cifras Sobre El Bienestar Subjetivo De Los Mexicanos." INEGI, Comunicación Social, Boletín de prensa, No. 431/12 (November 21). Available at: https://www.scribd.com/document/114103517/PRESENTA-EL-INEGI-CIFRAS -SOBRE-EL-BIENESTAR-SUBJETIVO-DE-LOS-MEXICANOS

INEGI. 2013. *Sistema de Clasificación Industrial de América del Norte.* Mexico: SCIAN 2013. Aguascalientes, Mexico: Instituto Nacional de Estadística y Geografía.

Instituto Federal Electoral. 2013. *La Calidad de la Ciudadanía en México.* México: IFE-Colmex.

Instituto Nacional de Estadística y Geografía. 2014. "México: nuevas estadísticas sobre informalidad laboral". Instituto Nacional de Estadística y Geografía. [Accessed 14 May 2015]. Available at: http://www.inegi.org.mx/est/contenidos/proyectos/ encuestas/hogares/regulares/enoe/default.aspx

International Labour Organization (ILO). 2012. *Global Wage Report 2012/2013: Is it the end of a low-wage production model in China?* [Accessed 15 June 2016]. Available at: http://www.ilo.org/global/about-the-ilo/newsroom/news/WCMS_192956/lang--en/ index.htm

Iriarte, Alicia, Mariana Vázquez and Claudia Bernazza. 2003. "Democracia y ciuda-danía: Reflexiones sobre la democracia y los procesos de democratización en América Latina", Col. Documentos Publicacion de l'institut internacional de gov-ernabilitat de Catalunya [Accessed 31 October 2016]. Available at: http://municipios .unq.edu.ar/modules/mislibros/archivos/dem_y%20ciudad.pdf

Kahneman, Daniel. 1999. "Objective Happiness," in D. Kahneman, E. Diener, and N. Schwarz (eds.), *Well-Being: Foundations of Hedonic Psychology*, pp. 3–25. New York: Russell Sage Foundation Press. [Accessed 4 July 2016]. Available at: http://profron .net/happiness/files/readings/Kahneman_ObjectiveHappiness.pdf

Kaku, Michio. 1997. *Visões de futuro—Como a ciência revolucionará a o século XXI.* Rio de Janeiro: Rocco.

Katz, Richard and Peter Mair. 2004. "El partido cartel. La transformación de los modelos de partidos y de la democracia de partidos," in *Zona Abierta*, No. 108/109: 9–39. [Accessed 12 June 2017]. Available at: http://escueladegobierno.corrientes.gov.ar/ assets/articulo_adjuntos/2710/original/Katz_y_Mair-El_partido_cartel .pdf?1484567739

Keck, Margaret E. 1992. *The Workers' Party and Democratization in Brazil.* New Haven, CT: Yale University Press.

Keynes, John M. 1930. "Economic Possibilities for Our Grandchildren." Available at: http://www.econ.yale.edu/smith/econ116a/keynes1.pdf (Scanned from John May-nard Keynes, Essays in Persuasion, New York: W.W. Norton & Co., 1963, pp. 358–373). [Accessed 2 May 2015].

Kim Jong Ho. 1998. "Los partidos políticos y el proceso de democratización en Corea del Sur," in *Desarrollo económico y democratización en Corea del Sur y el Noreste Asiático.* Córdova: Ed. Comunicarte, pp. 17–41.

Koo, Hagen. 1993. "The State, Minjung, and the Working Class in South Korea," in Hagen Koo (ed), *State and Society in Contemporary Korea*. Ithaca and London, Cornell Univerty Press.

Krokou, Danai. 2015. *Estrategias de entrada al mercado chino: Guía práctica para PYMES y empresarios*. Translated by María J. Manzano. Teaneck, N.J.: Babelcube Inc.

Krugman, Paul. 2012. *Detengamos esta crisis, Ya!* México, Editorial Paidós.

Lacey, Hugh. 1998. *Valores e atividade científica. Coleção Filosofia da ciência e epistemologia*. São Paulo: Discurso Editorial.

Lagarde, Christine. 2014. *Innovation, Technology and the 21st Century Global Economy*. International Monetary Fund. http://www.imf.org/external/np/speeches/2014/022514.htm [Accessed 4 April 2015].

Latinobarómetro. 2006. *Informe Latinobarómetro 2005: 1995–2005 Diez años de opinión pública*. Santiago de Chile: Corporación Latinobarómetro.

Latour, Bruno. 1992. *Ciencia en Acción—Como seguir a los científicos e ingenieros a través de la sociedad*. Barcelona: Editorial Labor.

Le Breton, David. 1995. *Antropología del cuerpo y modernidad*. Buenos Aires: Nueva Visión.

León, José Luis. 2005. "Corea del Sur: las transiciones múltiples de una economía posdesarrollista." Paper presented at the II Encuentro de Estudios Coreanos en América Latina, Colegio de México.

León, José Luis. 2006. "Autoritarismo y Democracia en Corea del Sur: Teoría y Realidad," in *Los intersticios de la democracia y el autoritarismo. Algunos casos de Asia, África y América Latina*. Romer, Cornejo, Buenos Aires, CLACSO, pp. 45–71 [Accessed 14 March 2014]. Available at: http://biblioteca.clacso.edu.ar/clacso/sur-sur/20100707072111/3_leon.pdf

Luhmann, Niklas. 1980. *Legitimação pelo procedimento*. Brasília: UNB.

Lykkergaard, John and Westergaard, Kurt. 2012. *Kurt Westergaard: The Man behind the Mohammed Cartoon*. Tilst, Denmark: Mine Erindringer (My Memoirs).

Malagón Plata, Luis Alberto. 2007. *Currículo y pertinencia en la educación superior*. Bogotá, Colombia: Cooperativa Editorial Magisterio.

Marchini, Genevieve. 2009. "Corea del Sur ante la crisis financiera global: costos vs. beneficios de la apertura financiera," in *Análisis México y la Cuenca del Pacífico*, Vol. 12, No. 36: 65–90.

Martínez, Juan Camilo. 2012. "La Asamblea Popular Nacional de China", *Observatorio Virtual Asia Pacífico*, Bogotá, 9 October 2012. [Accessed 1 June 2014]. Available at: http://asiapacifico.utadeo.edu.co/wp-content/uploads/2012/10/La-Asamblea -Nacional-Popular-de-China-_D_.pdf

McAfee, Andrew. 2013. "The Myth of Technological Unemployment." *Moneybox*, January 15. http://www.slate.com/blogs/moneybox/2013/01/15/the_myth_of_technological_unemployment.html [Accessed 30 December 2015].

McKenzie, David. 2013. "In China, 'Cancer Villages' a Reality of Life," CNN—Sección World—Part of Complete Coverage on China, 29 May. [Accessed 2 June 2014]. Available at: http://www.cnn.com/2013/05/28/world/asia/china-cancer-villages -mckenzie/

Merrill, Tim L. and Ramón Miró. 1996. *Mexico: A Country Study*. Washington: GPO for the Library of Congress. [Accessed 14 July 2016]. Available at: http://countrystudies .us/mexico/80.htm

MOFCOM (Ministerio de Comercio de la República Popular China). 2012a. "IX. The Election System—China's Political System," China Country Profile-Politics. [Accessed 29 May 2014]. Available at: http://www.china.org.cn/english/Political/26325 .htm2012b. "China's Current Legislation Structure—The Legislative System of China—China's Political System", China Country Profile-Politics (2012) [Accessed 29 May 2014]. Available at: http://english.mofcom.gov.cn/aarticle/zm/201205/ 20120508136310.html2012c. "I. The Constitutional System—China's Political System", China Country Profile-Politics [Accessed 29 May 2014]. Available at: http://www .china.org.cn/english/Political/26143.htm2012d. "III. The Party in Power—China's Political System", China Country Profile-Politics [Accessed 29 May 2014]. Available at: http://www.china.org.cn/english/Political/26151.htm 2012e. "Multi-party Cooperation and the Political Consultative System—China's Political System", China Country Profile-Politics [Accessed 29 May 2014]. Available at: http://english.mofcom.gov .cn/aarticle/zm/201205/20120508132561.html

MOFCOM (Ministerio de Comercio de la República Popular China). 2012f. "Communist Party of China—China's Political System", China Country Profile-Politics. [Accessed 29 May 2014]. Available at: http://english.mofcom.gov.cn/aarticle/zm/201205/ 20120508132541.html

NBS (National Bureau of Statistics of China). 2012. China Statistical Yearbook 2011. Beijing: China Statistics Press. Available at: http://www.stats.gov.cn/tjsj/ndsj/2011/ indexeh.htm

NBS (National Bureau of Statistics of China). 2014. China Statistical Yearbook 2014. Beijing: China Statistics Press. Available at: http://www.stats.gov.cn/tjsj/ndsj/2014/ indexeh.htm

NED (National Endowment for Democracy). 2008. *NED programs on Cuba*. National Endowment for Democracy. http://www.ned.org/about/about.html

Nietzsche, Friedrich. 1974 [1871]. *Obras Incompletas. Coleção Os Pensadores*. São Paulo: Abril Cultural, 1974. Primeira edição.

O'Donnell, Guillermo. 1982. *Burocrático Autoritario: Triunfos, derrotas y crisis*. Buenos Aires, Editorial de Belgrano.

O'Donnell, Guillermo. 1997. "Apuntes para una teoría del Estado", Documentos CEDES-CLACSO, No. 9.

O'Donnell, Guillermo and Schmitter, P. 1991. Transiciones desde un gobierno autoritario. Ed. Paidós. Buenos Aires.

OECD (Organisation for Economic Co-operation and Development). 2008. *OCDE Reviews of Innovation Policy. China*. Paris: OCDE.

OECD (Organisation for Economic Co-operation and Development). 2011. "Compendium of OECD Well-Being Indicators." Organization for Economic Co-operation and Development (OECD).

OECD (Organisation for Economic Co-operation and Development). 2013. "How's Life? 2013." OECD Publishing. Available at: http://www.oecd-ilibrary.org/economics/how -s-life-2013/the-oecd-better-life-initiative-concepts-and-indicators_how_life-2013 -5-en

Offe, Claus. 1994. *Contradicciones en el Estado de bienestar*. Madrid: Alianza Editorial.

Ogle, Jorge. 1991. *South Korea: Dissent within The Economic Miracle*. Atlantic Highlands, NJ: Zed Books.

OIT (International Labour Organisation). 2012. Organización Internacional del Trabajo: Informe mundial sobre salarios 2012–2013. http://www.ilo.org/global/research/ global-reports/global-wage-report/2012/WCMS_195244/lang--es/index.htm [Accessed 12 May 2015]. 2014. Organización Internacional del Trabajo: Informe mundial sobre salarios 2014/2015: Salarios y desigualdad de ingresos. http://www.ilo .org/wcmsp5/groups/public/---dgreports/---dcomm/---publ/documents/ publication/wcms_324818.pdf [Accessed 15 May 2015].

Olmedo García, Francisco. 2009. "GMOs to the Food Challenge." *Economic Mediterranean* 15: 123–141. [Accessed 15 March 2015]. Available at: www .publicacionescajamar.es/pdf/publicaciones-periodicas/mediterraneo -economico/15/15-257.pdf

Oubiña Falcón, Susana. 2015. "Los Chinos", Inventos e inventores—Sección Inventores (s/f I.E.S. Francisco Asorey). [Accessed 15 June 2017]. Available at: http://www.mjcv .es/susanaoubina/Inventos_Inventores/enlaces/chinos.html

Paley, Dawn. 2015. "Tlatlaya, Ayotzinapa, Apatzingán: State Terror in Mexico." *Opinion*. [Accessed 16 July 2016]. Available at: http://www.warscapes.com/opinion/tlatlaya -ayotzinapa-apatzing-n-state-terror-mexico

Panebianco, Angelo. 1993. *Modelos de Partido*. México, D.F.: Alianza Editorial.

Pardo, María and Nobila, María. 2000. *Globalización y nuevas tecnologías*. Buenos Aires: Editorial Biblos.

Pato, Miguel. 2011. "Antonio Garrido presenta en PD 'El lector de cadáveres'," *Periodista Digital*—Sección Ocio y Cultura, 27 October. [Accessed 23 August 2015]. Available at: http://www.periodistadigital.com/ocio-y-cultura/libros/2011/10/27/china -inventa-brujula-imprenta-barcos-insumergibles-europa-sumida-barbarie-lector -cadaveres-antonio-garrido-investigacion-criminal-csi-antiguo-muralla-china -occidente.shtml

Payne, Leigh A. 2000. *Uncivil Movements: The Armed Right Wing and Democracy in Latin America*. Baltimore and London: Johns Hopkins University Press.

Peiyan, Zeng. 2012. "The Establishment of the Socialist Market Economy," *Qiushi Journal* (English Edition), 4: 3 (July). [Accessed 1 March 2014]. Available at: http://english .qstheory.cn/leaders/201210/t20121010_185431.htm

Penoncello, Carlos A. 2005. "Luchas Estudiantiles, clases medias y democratización en Corea del Sur," pp. 253–269, in *Corea: una mirada desde Argentina*. Ed. by Eduardo D. Oviedo, Korea Foundation.

People's Daily. 2000. "Council of State," Section: State Structure, http://en.showchina .org/China/abc/01/200906/t338839.htm [Accessed 15 June 2016].

Piketty, Thomas. 2014. *El capital en el siglo XXI*. Mexico, Fondo de Cultura Económica.

Plastino, Angel Luis. 1993. "Relaciones entre el Estado y la universidad," in *Pensamiento Universitario*, No. 1: 46–48.

Prebisch, Raúl. 1950. *The Economic Development of Latin America and its Principal Problems*. New York: United Nations.

PricewaterhouseCoopers. 2015. "Technological Breakthroughs." http://www.pwc.co.uk/ issues/megatrends/technological-breakthroughs.html [Accessed 15 June 2017].

Quiroga, Águeda. 2003. "Introducción Al Análisis De Datos Reticulares. PráCticas Con Ucinet 6 Y Netdraw 1. Versión 2." Departamento de Ciencias Políticas, Universidad Pompeu Fabra. [Accessed 14 July 2016]. Available at: http://revista-redes.rediris.es/ webredes/talleres/redes.htm.

Rabotnikof, Nora. 2006. *En busca de un lugar común. El espacio público en la teoría política contemporánea*. México: Instituto de Investigaciones Filosóficas-UNAM.

Rainie, Lee and Wellman, Barry. 2012. *Networked: The New Social Operating System*. Cambridge: MIT Press.

Redacción Web. 2014. "BBC denuncia precarias condiciones laborales de los trabajadores de Apple", *La Voz de Michoacán*-Sección Mundo, 19 December. [Accessed 4 May 2015]. Available at: http://www.lavozdemichoacan.com.mx/bbc-denuncia-las -condiciones-laborales-de-los-trabajadores-de-apple/

Reinoso, José. 2013. "China enfrenta al ansia de libertad de sus ciudadanos", *El País*-Sección Internacional, 8 January. [Accessed 1 June 2014]. Available at: http:// internacional.elpais.com/internacional/2013/01/08/actualidad/1357633457_539562 .html

Rifkin, Jeremy. 1995. *The End of Work—The Decline of the Global Labor Force and the Dawn of the Post-Market Era*. New York, Tarcher/Putnam.

Rivera Vargas, María Isabel. 2011. "Innovation Systems Interactions and Technology Transfer and Assimilation for Industrial Development. The Cases of South Korea and Mexico," pp. 25–51, in *Theory and Practice of the Triple Helix System in Developing Countries: Issues and Challenges*, Mohammed Saad and Girma Zawdie, eds. New York, USA: Routledge.

Robin, Marie-Monique. 2004. *Escadrons de la mort, l'école française*. Paris: Éditions La Découverte.

Robinson, William I. 1996. *Promoting Polyarchy: Globalization, U.S. Intervention, and Hegemony*. Cambridge, UK: Cambridge University Press.

Rojas, Mariano. 2009 "Economía de la Felicidad: Hallazgos Relevantes Respecto al Ingreso y el Bienestar," *El Trimestre Económico* LXXVI, No. 303: 537–573.

Romero Castilla, Alfredo. 2001. "Corea del Sur: del 'milagro económico' a la era del FMI". *México y la Cuenca del Pacífico*. Vol. 7, No. 22: 7–21.

Romero Castilla, Alfredo. 2003. "Sociedad civil y regionalismo en el proceso de democratización de Corea del Sur", trabajo presentado 2º. Encuentro de Estudios Coreanos en América Latina, México [Accessed 28 November 2013]. Available at: http://www.uba.ar/ceca/actividades-conferencias.php

Romero Castilla, Alfredo. 2005. *El sistema político de Corea del Sur: Historia, Desarrollo Económico y democratización*. [Accessed 28 November 2013]. http://www.uba.ar/ceca/download/sistema-politico-de-corea-del-sur.pdf

Romero Castilla, Alfredo. 2012. *México y la República de Corea: Reflexiones en torno a sus 50 años de historia*. México y la Cuenca del Pacífico. Vol. 1, No. 2: 21–42.

Rousseau, Jean-Jacques. 2005 [1754]. *Discursos. Discurso sobre las ciencias y las artes. Discurso sobre el origen y los fundamentos de la desigualdad entre los hombres*. Buenos Aires: Losada.

Rubio Barrios, Julio, Tshipamba Ntumbua and Ramírez Alvarado Luis. 2013. "La Legislación como instrumento del Desarrollo de la Ciencia, Tecnología e Innovación: El caso de Corea del Sur," in *Enfoques: Ciencia Política y Administración Pública*, Santiago, Vol. XI, No. 19, pp. 19–35.

Sáez, Emmanuel and Zucman, Gabriel. 2014. "Wealth Inequality in the United States: Evidence from Capitalized Income Tax Data." NBER Working Paper Series. http://www.nber.org/papers/w20625 [Accessed 3 April 2015].

Samaniego, Norma. 2015. *La participación del trabajo en el ingreso nacional. El regreso de un tema olvidado. Serie Estudios y Perspectivas*. México: CEPAL.

Sánchez Jiménez, Arturo. 2014. "En México empresarios y académicos están desvinulados: investigador". *La Jornada*, 31 October. [Accessed 17 December 2014]. Available at: http://www.jornada.unam.mx/ultimas/2014/10/31/en-mexico-empresarios-y-academicos-estan-desvinulados-investigador-7549.html.

Santarrosa, Jorge. 2005. "Burocracia y tecnocracia en Corea del Sur 1961–1981," in *Corea: Una mirada desde Argentina*, Eduardo Oviedo (ed.), Universidad Nacional de Rosario and Korea Foundation, pp. 89–100.

Sen, Amartya. 1999. *Desarrollo y Libertad*. Argentina: Editorial Planeta Argentina.

SEP—CIDE. 2010. Encuesta Nacional de Vinculación en Empresas. Secretaría de Educación Pública—Centro de Investigación y Docencia Económicas.

Shiva, Vandana. 2005. *Earth Democracy: Justice, Sustainability, and Peace*. Cambridge, MA: South End Press.

Silvert, Jaime, Jorge Santarrosa and Francisco Bauer. 1997. "La relación del Capital y el Trabajo. El Estado y el Movimiento Obrero en Corea del Sur. Un análisis en Perspectiva Histórica", in Silbert Jaime I. *La República de Corea Hoy: economía, sociedad, relaciones internacionales*, Buenos Aires, Comunicarte Editorial, pp. 151–170.

Simondon, Gilbert. 2007. *El modo de existencia de los objetos técnicos*. Buenos Aires: Prometeo.

Sloterdijk, Peter. 2000. *Regras para o parque humano*. São Paulo: Estação Liberdade.

Sotelo Valencia, Adrián. 2014. "Latin America: Dependency and Super-Exploitation," in *Critical Sociology* 40: 539–549.

Stix, Gary. 2006. "Owning the Stuff of Life." *Scientific American* 294(2), 76–83

Story, Dale. 1983. "Industrial Elites in Mexico: Political Ideology and Influence," in *Journal of Interamerican Studies and World Affairs* 25(3): 351–376.

Streeck, Wolfgang. 2011. "La crisis del capitalismo democrático," in *New Left Review*, No. 71 (November–December): 5–26.

Tapia, Luis. 2011. *El estado de derecho como tiranía*. La Paz, Bolivia: UMSA-Autodeterminación.

TeleSur. 2015. "Mexican Farmworkers Demand Attention after Police Violence," TeleSur, 11 May. [Accessed 16 June 2016]. Available at: http://www.telesurtv.net/english/news/Mexican-Farmworkers-Demand-Attention-after-Police-Violence-20150511-0004.html

Téllez Valdez, Julio. 2004. *Derecho informático*. Mexico: Editorial McGraw-Hill.

Tilly, Charles. 2007. *Democracia*. Madrid: Akal.

Toussaint, Eric. 2006. "Corea del Sur. El milagro desenmascarado." Conference held in Caracas, Venezuela, 14 November 2006. [Accessed 5 May 2014]. Available at: http://cadtm.org/Corea-del-Sur-el-milagro

Transparencia International. 2006. *Transparency International's Global Corruption Barometer 2006 Report*. Berlin, Germany: Transparency International. International Secretariat.

Turner, Jennifer. 2013. "Estados Unidos: El uso de la fuerza contra manifestantes para reprimir la libertad de expresión en Puerto Rico," in *Recuperen las calles: Represión y criminalización de la protesta en el mundo*, International Network of Civil Liberties Organizations (INCLO), pp. 4–9. http://www.cels.org.ar/common/documentos/INCLOProtestaSocial-Espanol.pdf [Accessed 25 March 2015].

Turnley, Jonathan. 2012. "Diez razones por las que Estados Unidos ya no es la tierra de la libertad," *Contrainjerencia*, February 3. http://www.contrainjerencia.com/?p=37657. [Accessed 3 April 2015].

UACP-UAZ. 2014. Encuesta Sobre Vinculación de la Unidad Académica de Ciencia Política de la UAZ. Zacatecas, México: Unidad Académica de Ciencia Política de la Universidad Autónoma de Zacatecas.

UIS (UNESCO Institute of Statistics). 2015. "Human Resources in Research and Development—Total R&D Personal", Browse by Theme—Science, Technology and Innovation. [Accessed 27 August 2015]. Available at: http://www.uis.unesco.org/DataCentre/Pages/BrowseScience.aspx

UNCTAD (United Nations Conference on Trade and Development). 2015a. "Foreign Direct Investment—Inward and Outward Foreign Direct Investment Flows, Annual, 1970–2013", UNCTADstat—Sección Statistics [Accessed 27 August 2015]. Available at: http://unctadstat.unctad.org/wds/TableViewer/tableView.aspx

UNCTAD (United Nations Conference on Trade and Development). 2015b. World Investment Report 2015. New York and Geneva: United Nations.

UNESCO (United Nations Educational, Scientific and Cultural Organization). 2010. *UNESCO Science Report 2010*. Paris: UNESCO.

Valencia Lomelí, Enrique. 2001. "Democratización y Crisis financiera: los desafíos de una transición herida. El caso de Corea del Sur y las secuelas de su crisis financiera en 1997–1998." *Mexico, Espiral* Vol. VII, No. 20, 91–133.

Veenhoven, Ruut. 1997. "Advances in Understanding Happiness." Revue Québécoise de Psychologie 18: 29–74. [Accessed 6 July 2016]. Available at: http://repub.eur.nl/pub/16324/97c-full.pdf

Williamson, John. 2002. *Did the Washington Consensus Fail?* Washington, DC: Peterson Institute for International Economics.

Winer, Sonia. 2005. "El rol de los movimientos sociales en Corea del Sur." Presented 8 June at Corea, una mirada desde Argentina, sponsored by the Universidad del Rosario, Argentina.

World Justice Project. 2015. *Rule of Law Index 2014*. Washington, D.C.: World Justice Project.

Worldometers. 2015. "China—Population by Country," Population, *Dadax*. [Accessed 23 July 2015]. Available at: http://www.worldometers.info/world-population/china-population/

Xing, Yuqing. 2012. "The People's Republic of China's High Tech Exports: Myth and Reality," *ADBI Working Paper Series*, Asian Development Bank Institute, No. 357 (April).

Xing, Yuqing and Neil Detert. 2010. "How the iPhone Widens the United States Trade Deficit with the People's Republic of China," *ADBI Working Paper Series*, Asian Development Bank Institute, No. 257 (December).

Index

CPSIA information can be obtained
at www.ICGtesting.com
Printed in the USA
LVHW01s1236280918
591634LV00004B/7/P